BEEKEEPING STUDY NOTES

for the

BBKA CERTIFICATE
IN
BEEKEEPING HUSBANDRY

Prepared by:

J.D.YATES B.Sc.(Hons), C.Eng., FIEE.
and
B.D.YATES SRN, SCM.

BEE BOOKS NEW AND OLD
10, Quay Road,
Charlestown,
Nr. St.Austell
Cornwall PL25 3NX

First published 2002

Printed and bound in Great Britain by
Penwell Limited, Station Road, Kelly Bray, Callington, Cornwall PL17 8ER

PREFACE

With the advent of this new BBKA examination namely, Cert: have been asked to prepare notes for those Candidates wis examination. We have adopted the same format as our t examinations. Accordingly, we address every section of the s most likely points that an examiner may reasonably be expected to raise during examination.

The examination is an oral one combined with practical manipulation of honeybee colonies in the Candidate's apiary. It has evolved for those who consider that written examinations are not their forté. We have heard it said that this examination is the soft option instead of gaining the Intermediate Certificate via the written route. We hope that in the fullness of time this proves to be a myth.

The contents section of our notes is the syllabus for the Examination and has the same numbering as the official BBKA syllabus. We have incuded as appendices the Prospectus, BBKA notes on the Record Book, Record Keeping together with Application to Enter and Suggested Reading Texts.

It is necessary to have passed the Basic and to have kept and managed bees for at least three years. This seems to us to be too short a minimum experience with bees but time will tell. As the syllabus was first published in April 2000 there is a paucity of experience with this new examination to decide whether the parameters and requirements are meeting the objectives. We hope that these notes will indeed be part of "a route to Beekeeping Excellence" and will see prospective candidates through the examination.

Finally, we express our thanks to Mr Peter Wearden, the Divisional Environmental Health Officer of South Hams District Council, for checking those parts of the manuscript dealing with food hygiene and nuisance caused by bees and also to John Annett of the Harrogate and Ripon Beekeepers' Association for his diligent proof reading.

JD & BD Yates,
Newton Ferrers, Devon. 2001

CONTENTS

SYLLABUS

General

1.1 The Assessment will take place at the Candidate's apiary.

1.2 The practical assessment will be conducted on any suitable day agreed with the Assessor in May, June or July. A minimum of three of the Candidate's own colonies shall be made available for examination. It is desirable that the Candidate's colonies are good tempered and do not follow more than about 5 metres from their hive.

1.3 In addition to three colonies, there shall be a queen-right nucleus for developing into a colony, or alternatively a mating nucleus used in conjunction with a queen rearing procedure.

1.4 The Candidate shall have in hand at the time of the assessment, a queen rearing procedure underway to demonstrate the Candidate's ability to rear queens suitable for the needs of the apiary.

1.5 On the day of the Assessment the Candidate's Record Book and any individual hive records shall be made available to the Assessor.

1.6 The Assessor will look for satisfactory use of the smoker and proficient manipulation of bees during the required demonstrations.

1.7 The Candidate will be able to describe the more usual situations that may result in honeybees becoming a nuisance to the public or livestock.

1.8 The Candidate will discuss methods of beekeeping and how these have been influenced by local conditions. Reference will be made to the choice of hives from those types commonly in use in the area, the merits of top and bottom bee space and apiary equipment.

1.9 The Candidate will describe the procedures for general maintenance including preservation of hives, fumigation of comb and equipment, prevention of wax moth damage, the use of predator guards, storing combs and general apiary hygiene.

1.10 The Candidate will describe the associated dangers of robbing and describe the methods in place to prevent robbing and to end robbing once started.

1.11 The Candidate will describe the methods used to minimise drifting and circumstances when diverting bees to another colony can be an advantage.

Practical Beekeeping

The Candidate will be able to demonstrate understanding of, and ability to perform the following tasks:

2.1 Describe the role of good hygiene in the apiary. The Candidate will demonstrate satisfactory procedures, both with personal effects and apiary equipment.

2.2 Describe the appearance of healthy brood and, in contrast, the appearance of larvae, brood pattern and cell capping that will require further investigation.

2.3 Describe the procedures taken to avoid or reduce the transmission of infectious diseases and

demonstrate that these are followed. Spare equipment used by the Candidate will be examined.

2.4 List the reasons for comb renewal and demonstrate the procedures adopted.

2.5 Discuss the progress of the colonies as described in the Record Book and the intentions for the rest of the season.

2.6 Review the age of existing queens and plans for their replacement. Describe how replacement of queens is carried out.

2.7 Describe the methods taken throughout the year to monitor and control varroa to non-damaging levels. Demonstrate the use of varroa control equipment in the apiary. Examine a brood chamber and floor for varroa. Demonstrate the use of comb for trapping mites in drone cells.

2.8 Describe the routine measures taken to look for disease in the colony.

2.9 Demonstrate the inspection of a brood comb for brood diseases.

2.10 Demonstrate taking a sample for the diagnosis of adult bee diseases.

2.11 Describe the factors that may initiate swarming and the indications that a colony is making preparations to swarm. Describe the economic and social effects of swarming and the procedures that are used to control swarming. Describe the procedures for creating an artificial swarm or any other method that may be used to ensure a colony does not swarm.

2.12 Describe the procedures used up to the time of the assessment in the queen rearing method demonstrated and explain what has yet to be done. Describe what is intended for the queens that have successfully mated. Describe the procedure that will be adopted to introduce queens into a colony.

2.13 Demonstrate marking and clipping a queen, or use a drone as a substitute if appropriate.

2.14 Describe the advantages of marking and clipping queens.

2.15 Describe the procedures adopted for adding supers.

2.16 Describe the procedures adopted when removing supers for honey extraction.

2.17 Describe how combs and cappings are dealt with after extraction. Refer to the methods adopted for clearing bees from supers and any treatment of the supers and combs that is routinely carried out before storage.

2.18 Describe how the colonies are prepared for winter and the timing of carrying out these arrangements.

2.19 Describe methods and reasons for feeding sugar syrup, candy, pollen and pollen substitute.

2.20 Describe how super combs are stored and the measures taken to combat wax moths.

2.21 Discuss the influence of honey production on apiary procedures.

2.22 Demonstrate how to prepare a colony for moving to another apiary.

2.23 Describe procedures used for moving a colony a short distance within an apiary and to another site beyond normal flying distance, making reference to the difficulties and dangers involved.

2.24 Describe the procedures used to prepare a nucleus colony. Discuss the many uses for a nucleus colony.

2.25 Describe the actions required to deal with a vicious stock of bees.

2.26 Demonstrate the procedures used for uniting two colonies and the precautions that need to be taken.

2.27 Demonstrate how beeswax is recovered with reference to the actual equipment used.

Natural History and Behaviour

The Candidate will be able to describe the following and explain their relevance to practical beekeeping:

3.1 The different races of honeybees and their characteristics.

3.2 The main external features of the drone and the two female castes.

3.3 The function of the hypopharyngeal glands, the Nasonov gland, the wax glands, the alimentary canal and the sting of the worker and queen.

3.4 The factors in the production of brood, which result in workers, drones and queens.

3.5 The mating of drones and queen.

3.6 The main stages in the development of the brood from egg to emerging adult and also the life expectancy of workers, drones and queens.

3.7 The changing circumstances throughout a year that influences the egg laying of a queen, indicating how the numbers will vary.

3.8 The nutritional requirements of honeybees and their main sources.

3.9 The signs in a colony of a drone laying queen and laying workers. Explain how these may arise and how they may be dealt with.

3.10 The seasonal variation in the hive population during a year including survival behaviour in winter.

3.11 The effect of weather on a summer colony and foraging.

3.12 The type of work done by a worker honeybee throughout its life including reference to summer and winter bees.

3.13 The collection of nectar and how it is converted into honey suitable for storing in sealed comb.

3.14 The collection of pollen and how it is carried to the colony and used.

3.15 The production of wax and how it is used in the colony.

3.16 The collection of water and propolis and how they used in the hive.

3.17 The factors that may give rise to swarm, supersedure and emergency queen cells.

3.18 The use made by honeybees of the alarm pheromones and the effect these have on the way bees are managed.

Foraging

The Candidate will be able to demonstrate understanding of:

4.1 The main plants of local importance to the bees throughout the year, giving details of flowering times.

4.2 Any measures taken by the Candidate to enable the bees to forage on a particular crop and any special action needed as a result of foraging on local crops or a crop to which bees have been taken. Rape, Heather and Borage are three possible examples.

4.3 Honeydew, being able to name sources and describe the impact of honeydew in the area of the Candidate.

4.4 Any sources of undesirable nectar found in the locality of the Candidate.

Disease, Pests and Poisoning

The Candidate will be able to:

5.1 Describe how disease can be spread between colonies and how good management practice can reduce disease occurring.

5.2 Describe the signs of American Foulbrood.

5.3 Describe the signs of European Foulbrood.

5.4 Describe what actions shall be taken to comply with statutory requirements if a brood disease is suspected.

5.5 Describe how to distinguish between female Varroa and female Braula coeca.

5.6 Describe the method adopted in the Candidate's apiary to monitor and control Varroosis.

5.7 Discuss the impact of virus damage related to Varroosis.

5.8 Discuss the impact of re-infestation of Varroa on the management and timing of Varroa control.

5.9 Describe the impact of Nosema disease on a honeybee colony, and its diagnosis and treatment.

5.10 Describe Acarine, its detection and recommended method(s) of control.

5.11 Describe Chalk Brood and Sac Brood, detection and control measures.

5.12 Describe what action can be taken by beekeepers to avoid damage to honeybees by spraying.

5.13 Describe the signs that suggest a case of poisoning. Describe the actions that should be taken. Describe how a sample of affected bees is collected, packaged and labelled and where this is sent.

Honey and Honey Processing

The Candidate will be able to:

6.1 Demonstrate the apiary equipment normally used specifically for the production of honey, including super comb frames and spacers; section apparatus; queen excluders, devices for clearing bees from supers. Discuss their use.

6.2 Discuss the preparation of colonies for specific nectar flows in the area.

6.3 Demonstrate the equipment used to extract and prepare the honey produced in the apiary and show the place used for processing and packing honey.

6.4 Have available for inspection by the Assessor, typical samples of packed honey ready for the table and for retail sale.

6.5 Describe the arrangements made by the Candidate for extracting honey from the comb and the preparation of Comb Honey.

6.6 Describe the processing and storage arrangements for the honey and packaging for sale.

6.7 Describe how the requirements for public health and safety, consumer protection, food hygiene, as overseen by the Environmental Health Officer, apply to Candidates in the area.

6.8 Demonstrate familiarity with current regulations and any other statutory requirements as they affect those offering honey for sale.

6.9 Describe Liquid Honey and Set Honey (both granulated and soft set) and a method that may be employed to obtain these with good quality results, including mention of the recommended temperatures for satisfactory results.

6.10 Describe the spoilage of honey particularly by fermentation (including the effect of water content, storage temperature and the presence of yeast).

Stings

The Candidate will be able to:

7.1 Describe how to deal with a person who has been stung by a bee but shows no effect other than discomfort and slight local swelling.

7.2 Describe precisely the action to take when a person who has been stung by a bee, exhibits a severe reaction or anaphylactic shock.

Appendices

1. Prospectus for the examination.
2. Record Book as recommended by the BBKA.
3. Record Keeping as recommended by the BBKA.
4. Application to enter the Examination.
5. Reading Texts as recommended by the BBKA.
6. Colony development during the active season.
7. Migration and Evolution of the honeybee.
8. Average population cycle.
9. The recovery position.

** ** ** **

FOREWORD

John and Dawn Yates have already done great service to beekeepers studying for the BBKA theory exams, with their green & orange book study notes covering the eight modules. There can be few examination minded beekeepers who have not benefited from these notes when tackling the BBKA theory modules for the Intermediate and Advanced Certificates.

The ubiquity of their study notes lies in presenting knowledge in the same specific objective format that the syllabi were written in. For some detractors therein lies a weakness in that the student is exposed to no other learning resource. My own view is that any mechanism that makes initial learning easier by presenting most of the relevant information in one place must be better than having to individually search and research each topic, with all the confusion of conflicting opinions at their unsure and early stage of learning. Those who merely want the certificate will in any case look no further than these notes. Those who seek understanding will go on to expand their knowledge by exploring further through other sources and resources.

It is those people who want to take the study further that interest me most from my position as Examiner and Course Director for the National Diploma in Beekeeping Examination Board. The usual entry qualification for taking the NDB exam is the BBKA Advanced Certificate plus the Senior Practical exam. In itself the old Senior Practical stood outside the modular system to provide a qualification route for the practical beekeepers - in bee husbandry. In recent years a number of candidates for the NDB exam have found difficulty with the bee husbandry skills whilst yet excelling in the theory exams. In itself a telling compliment to the BBKA theory module study system, but highlighting the need for practical beekeeping skills development. To help remedy this deficit, the NDB Board initiated the new, residential week long Advanced Bee Husbandry Course, first run in 2000 at the National Bee Unit and to be repeated every second year.

The BBKA Examination Board were by no means idle in this concern and following proposals from some Avon beekeepers set about the development of new practical examinations in bee husbandry, culminating in the BBKA Certificate in Bee Husbandry for which these study notes have been prepared by John and Dawn.

Once again the attention to detail and the well researched material is evident. I am sure that these notes will prove to be of great benefit in helping to develop good bee husbandry skills.

K. Basterfield.
B Sc(Hons), C Eng, MIEE, NDB

GENERAL

1.1 The Assessment will take place at the Candidate's apiary.

It seems curious that this has been included in the syllabus for the examination when it has been included in the prospectus. We must assume that because it is included in the syllabus then the Assessor may wish to discuss with the Candidate various aspects of the apiary in a general manner including such items as why the site was selected, its efficacy as a site in the general area and its juxtaposition with neighbours, livestock and other beekeepers. Therefore, with this premise in mind, we append below the general considerations and the detailed considerations for siting colonies in a home apiary.

1.1.1 General considerations.

Before siting colonies in an apiary, consideration should be given first to the site itself and indeed whether it is suitable for keeping bees. Criteria for the site are as follows:

> a) Is there adequate forage in the surrounding area for the colony to support itself and can it readily obtain water throughout the year?
> b) There must be no question of danger to humans, particularly children, or animals. One sting can kill an allergic subject if not given expert medical treatment very quickly.
> c) Ideally, the apiary should be in a place where nobody except the beekeeper can be stung. This criterion is nigh impossible to achieve and siting of the colonies in the apiary becomes of vital importance to minimise this risk.
> d) Under no circumstances should an apiary be established adjacent to a public thoroughfare, even if there is a barrier (eg. hedge or wall) of suitable height between. Bad tempered bees while being manipulated will attack moving human and animal targets at quite large distances from their hive.
> e) The site should not be in a frost pocket and protection from the prevailing winds is most desirable.
> f) The site should be free from any form of flooding and not under trees in or on the edge of a wood.
> g) The site should be accessible by road at all times of the year.
> h) The site should be surrounded by a stock-proof fence if it is adjacent to pasture where livestock is likely to be grazing.
> i) Finally a matter of ethics relating to out apiary sites which are often on a farmer's land; if another beekeeper has bees close by on the same owner's land, then look elsewhere for another site no matter what the farmer says.

It cannot be over emphasised that the utmost care must be taken in siting an apiary and the colonies in it. Bees cannot be moved around like other livestock. If there is any doubt about any aspect of siting, expert advice should be sought.

1.1.2 Detailed considerations for siting colonies in a small home apiary.

A small home apiary implies that there is a dwelling house nearby with other people, children and domestic animals, if not in the in the same grounds then on neighbouring property. One might reasonably ask what is small? The answer is somewhat subjective but we would have a stab at 5 or 6 maximum stocks of bees.

The same or similar sting considerations apply to an out apiary, therefore the following points require attention:

a) Stocks must be sited so that the flight path of the bees avoid footpaths and areas where there is likely to be any human or animal activity. Stocks can be sited so that bees have to fly up and over hedges and fences thereby getting the bees to a safe height above anyone on the ground. Under normal circumstances such an arrangement is quite workable but aggressive bees must be considered. It is absolutely essential to have a 'bolt hole' prepared over 3 miles away, so that a stock may be moved in an emergency.
b) There must be plenty of space around each stock for colony manipulations and maintaining the site (eg. grass cutting). A distance of 6ft. between colonies would not be out of the way for setting up nucs and doing artificial swarms between the adjacent stocks.
c) Space should be allowed at the planning stage for expansion in the future; this aspect is often overlooked.
d) The layout of the stocks should be in an irregular fashion in order to minimise drifting.
e) Hives should be provided with permanent bases to raise the floorboard off the ground to prevent damp and possible rot starting to occur in the lower woodwork of the hive. Concrete bases are undoubtedly the best but try a temporary solution until the site has been tried out for a couple of years.
f) The height of the top of the brood chamber, to minimise too much bending, is very important if large numbers of stocks are involved. Even ½ dozen hives become a real pleasure if they are at the right height compared with being too low. A point to consider when designing hive stands.
g) In home apiaries it is best to site the stocks out of sight of neighbours if this can be done.
h) Bees in the stocks will at some time swarm despite the best efforts of the beekeeper to prevent this happening. Shrubs and trees around the stocks are useful for the swarms to hang on.
i) A certain amount of shade from nearby trees is useful particularly at mid-day during the summer. Stocks should not be sited under trees where rain drops can fall from them onto the hive in winter; it disturbs the colony.

Provision should be made for storing spare equipment near to the apiary, preferably a discrete shed for the multitude of bits and pieces. There is a risk to life, albeit fairly remote, wherever bees are kept and if this is remembered when planning an apiary, the chances of success are well assured. Having said all this, we believe that suburban gardens are becoming so small and houses so close together that they are unsuitable sites for keeping bees and certainly not the place for beginners.

1.2 The practical assessment will be conducted on any suitable day agreed with the Assessor in May, June or July. A minimum of three of the Candidate's own colonies shall be made available for examination. It is desirable that the Candidate's colonies are good tempered and do not follow more than about 5 metres from their hive.

The first sentence is again part of the prospectus and requires no further comment except that the Candidate should be prepared to do the examination under adverse weather conditions. The Assessor may have travelled a long way only to find the weather inclement and he may wish to press on.

The inference in the second sentence is that the Candidate shall select 3 colonies for presentation to the Assessor for the examination and the final sentence indicates that good temper is required because a large part of the examination is likely to be manipulating all three colonies.

We postulated a simple method of measuring the aggression of colonies in an apiary when we published our Study Notes for Module 1 in 1996 which is worth quoting and is as follows:

> "We are all aware that it is possible to open up some colonies of bees without a veil or other protection and it is unlikely that a sting will be suffered by the operator, assuming that he knows what he is doing and handles his bees well. We are also aware that other colonies are virtually impossible to handle. In between these two extremes is the wide variation of temperament. The best measurement to assess the defensive nature of the bees is, in our opinion, to take note of the followers. If, after manipulating a colony, there are no followers 2 metres from the back or side of the hive then it is a suitable bee for keeping in a suburban garden. At 2 metres one should be able to remove the veil and not get pestered by a guard bee or follower.

> If bees follow up to 5 metres then, in our opinion, they are unsuitable for a suburban garden. Such bees will need careful handling during a manipulation and would only be suitable in an isolated out apiary.

> If bees follow beyond 5 metres then we would consider requeening them in favour of a more favourable strain."

To comply with the syllabus, using our yardstick which appears to have been adopted by the BBKA, if your bees follow up to 5 metres then the apiary should be an isolated out apiary.

1.3 In addition to three colonies, there shall be a queen-right nucleus for developing into a colony, or alternatively a mating nucleus used in conjunction with a queen rearing procedure.

Note that the syllabus states "shall be" and not 'will be'. The rules for the use of 'will' and 'shall' are often ignored in English. 'Shall' is mandatory whereas 'will' is a future tense when combined with a verb. It is, therefore, mandatory to have a nucleus in the apiary, as stated in the syllabus above, on the day of the examination otherwise a failure should (not could!) be guaranteed. If the assessment is to last about 2 hours there is unlikely to be sufficient time to start making one while the Assessor is present.

For making up nuclei see also Section 2.24.

1.4 The Candidate shall have in hand at the time of the assessment, a queen rearing procedure under way to demonstrate the Candidate's ability to rear queens suitable for the needs of the apiary.

Note again that this is a mandatory requirement to pass the examination because of the word "shall". The sting in the tail here is, in our opinion and no pun intended, "suitable for the needs of the apiary". We believe that the Candidate should be able to convince the Assessor that the offspring of the reared queens will be suitably docile for use in that particular apiary.

For queen rearing methods reference should be made to Section 2.12.

1.5 On the day of the Assessment the Candidate's Record Book and any individual hive records shall be made available to the Assessor.

Here we have another mandatory requirement. In respect of hive records; if the assessment is in mid-summer then hive record cards for the previous season would be required particularly if the colony has been used for queen rearing. BBKA leaflet entitled "Maintaining Beekeeping Records" should be obtained as this leaflet makes specific reference to this examination. One word of caution, the leaflet mentions "Room" and then goes on about the available space for stores; this should, in our opinion, be space for bees. Supers are primarily for bees; consider it a bonus if they are filled with honey.

1.6 The Assessor will look for satisfactory use of the smoker and proficient manipulation of bees during the required demonstrations.

As this examination is an assessment in beekeeping husbandry we believe that this particular syllabus item should be made mandatory. It is not addressed anywhere else in the syllabus and it is vital that a firm understanding of how a colony should be manipulated has become second nature to the Candidate.

The proficient manipulation of a honeybee colony requires the proficient use of the smoker. A colony can only remain under the control of the beekeeper if:

- The bees are in the hive.
- The colony has been subdued with smoke.

1.6.1 When opening a colony for, say, a routine inspection the following sequence should be followed:

a) Approach the colony quietly and gently place any spare equipment on the ground near the hive.
b) Observe the normal activity at the entrance.
c) Smoke at the entrance (6 good puffs of the smoker). Don't be timid with the smoker at this

stage.

d) Wait for at least two minutes (you should know why).

e) Gently remove the roof and place it on the ground, upside down, behind the hive with a corner pointing to the back of the hive. This may sound a bit pernickety but it is now all ready in the right position to put the supers on.

f) Now study the record card, check that you have all the equipment required and that you are sure what has to be done.

g) Two small puffs of smoke at the entrance; we're coming in now!

h) Ease the supers up on one side with the hive tool, tilt and smoke well between the supers and the queen excluder.

i) Lift off supers and place gently on the upturned roof. No bees will be squashed as there are only 8 small points of contact.

j) Smoke the bees down so that the queen excluder is clear of bees and then gently lever and twist it off. Hold it over the hive and shake any remaining bees off and place behind the hive.

k) NOW PUT THE SMOKER BETWEEN YOUR KNEES ready for use.

l) Remove the dummy board, shake off any bees into the top of the hive with a sharp tap and lean it up at the rear of the brood chamber with the queen excluder.

m) A small puff of smoke to clear the bees at the lugs of the first frame (where you want to hold them) and gently remove it, inspect and replace it hard up against the side of the brood chamber.

n) Don't forget to glance at the entrance.

o) If the bees are tending to well up and fly off, smoke them down.

p) Repeat the operation with every frame always keeping the bees down in the hive with the smoker. No jerking or bumping or fast movements are acceptable.

q) Replace the dummy at the opposite end of the brood chamber and lever the frames and dummy with the hive tool so that there are no gaps between the spacing devices. This will prevent a build up of propolis which would change the frame spacing.

r) Smoke the tops of the frames clear of bees and replace the queen excluder.

s) Replace the supers carefully without squashing any bees.

Note: if there are no supers on the colony, check that the queen is not on the underside of the crown board when it is removed.

1.6.2 If the bees are difficult to keep down in the hive with smoke then use cover cloths, but avoid them if possible. We have found that sunlight/daylight also tends to subdue some strains of bee. If cover cloths are needed most of the time then, in our opinion, the strain of bee could usefully be changed.

1.6.3 There is an acquired art in the use of smoke; watch carefully a competent beekeeper at work, the art is easy to learn.

1.6.4 If you feel the colony is getting out of control, then close it down before you get into a 'tail spin'. It is the most sensible thing to do and is good beekeeping practice, even during an examination with the permission of the Assessor.

1.6.5 Demonstrate the use of the smoker.

1.6.5.1 During the examination, the part where the Candidate has to manipulate the colony, the Assessor will be noting how you use the smoker. First you will be asked to light it. A seemingly simple task but Candidates have been failed for not being able to accomplish the lighting procedure. It is no good filling the smoker with fuel and throwing a lighted match on top; it will go out no matter how hard the bellows are worked. The burning fuel must be at the bottom of the fire box. You should be able to light the smoker in the apiary with a breeze blowing. We believe that using a small gas blow torch to be a doubtful starter; no pun intended!

1.6.5.2 It is important that the smoker is burning well, full of fuel and producing plenty of cool smoke before approaching the hive. To achieve this state of affairs will take a few minutes to produce some hot ashes at the bottom of the fire box necessary to keep the smoker burning by itself without pumping the bellows. The Assessor will be watching closely as it is unacceptable for the smoker to go out at the beginning of the hive manipulation. If the smoker has been well lit it will continue to burn slowly by itself if left in the upright position and produce an adequate supply of smoke when the bellows are worked.

1.6.5.3 See above on the use of the smoker during the manipulation. We cannot stress too strongly that there is an art using the smoker which is easy to learn. Do observe the Master Beekeepers that you come across while they are working their colonies; you will learn a lot.

1.6.5.4 The Assessor will still be watching what you do with your smoker when the manipulation is complete, the hive closed up and the smoker is no longer required. Plug the nozzle with some green grass or better still have a wooden plug that fits snugly in the nozzle tied to the smoker permanently with a piece of string. On no account should the hot ashes be emptied out of the smoker, as they could be a fire hazard.

1.6.5.5 It will be noted that we recommend parking the smoker between your knees during a manipulation. We have met the odd beekeeper who has not felt comfortable using this method and prefers a hook on the rear of the bellows to hang it on the side of the hive. This is OK if cover cloths are not being used.

Finally, it should be noted that some bees do not react favourably to smoke and all bees will not react to smoke if there are no liquid stores for them to gorge on.

1.7 The Candidate will be able to describe the more usual situations that may result in honeybees becoming a nuisance to the public or livestock.

Nuisance is caused in a variety of ways but the three most important are:
  • Stinging.
  • Swarming.
  • Fouling newly washed clothes put out to dry and soiling parked cars.

1.7.1 Stinging.

When neighbours are stung it is generally as a result of the bees being of doubtful temperament.

1.7.1.1 How the variable temperament of bees arises.

In order to address this part of the syllabus it is necessary to have a clear understanding of why the temperaments of our honeybees are variable and how this arose. There is a very wide range of temperaments from the very docile to the very aggressive or, put another way, from those that exhibit little defensive instinct to those with a very strong defensive instinct, whether the colony is disturbed or remains undisturbed.

By various scientific techniques, including palaeontology, it is generally agreed that the origin of the honeybee was in that part of Africa that is now known as Kenya somewhere between 20 and 30 million years ago. This was before India split off from Africa and before the Red Sea and Rift Valley were formed. The land mass now called India carried with it some of the earlier bee-like insects which developed into the Eastern species of bees, ie. *Apis dorsata*, *A. cerana* and *A. florea*.

Reference now to the diagram of the 'Migration of the Honeybee' in Appendix 7 shows the origin in Kenya and the main migration routes marked with double line arrows northwards, southwards and westwards. Major races developed notably *A.m.capensis* in the Cape of Good Hope area, *A.m.adansonii* in the equatorial strip, *A.m. fasciata* (the Egyptian bee) in the north east and *A.m.intermissa* (the Tellian or Arab bee) in the north west.

The Tellian bee is considered to be a major race from which many other strains have developed. In its native part of northwest Africa, it developed in a very hostile environment and presumably adapted to these conditions by having to defend its colonies against determined predators eg. hornets, etc. It migrated northwards before the last melting of the ice cap c. 10,000 years ago and established the well known races in NW Europe such as the Iberian bee, the French Black bee and the English bee. It migrated as far north as Finland in latitude 60°N.

Similarly, the other major race (*fasciata*) migrated northwards also providing the Italian bees (*ligustica*), Greek, Caucasian, Carniolan, etc. After the ice cap melted the UK was cut off from the continent and the Mediterranean was formed, isolating Europe from Africa. At that time we had a discrete race of bee in the UK now known as the English Black bee or the English Brown bee. It was decimated by disease in the early 1900s (n.b. Isle of Wight Disease or Acarine) and the Government of the day offered subsidies to beekeepers to import bees from the Continent. The result was that most of the races in Europe were imported and over the last 50 to 80 years these races have interbred and we now have a hotchpotch of mongrels with a very mixed pool of genes.

All the bees with their origins emanating from the Tellian bee have to a greater or lesser degree the defensive instinct, while other bees from Italy, Greece, etc. have a very weak defensive trait. Breeding can only be true with pure strains, while breeding with mongrels is known to be an erratic and unpredictable procedure.

Herein then lies the root of our problem in the UK. Using mongrel bees and taking pot luck with the matings will result in a wide range of temperaments in the offspring. The only beekeeper who claims experience with the old English bee, and has committed his findings to paper, is Bro.Adam. He considered the bee had some very desirable features but also some very bad ones, notably bad temper. BIBBA claim that there are pockets of the old English bee with a good temper and campaign for its breeding and use in the UK.

Before leaving this introduction to understanding the temperament of our bees it must be pointed out that other characteristics will also be present in mongrels, such as high tendency to swarm, heavy propolis gatherers, early starters/late finishers, longevity, economy on stores for overwintering, resistance to different diseases, etc. We know from experience that it is possible to eliminate bad temper fairly quickly by culling those queens producing bad tempered bees. However, it is nigh impossible to control more than two variables without recourse to specialised isolation and mating techniques. Our own breeding programme concentrates on good temper and minimum swarming tendency which can be achieved by any beekeepers who put their minds to it. Unfortunately, it is a very small minority who control the temperament of the bees they keep and we put this down to two traits that have developed over the last 50 years as follows:

a) Newcomers have been encouraged to wear very adequate protective clothing which has become available, making them 'safe' from stings under most conditions.
b) Few associations encourage their membership to manipulate their bees without gloves.

1.7.1.2 The defensive qualities of different strains.

We mentioned our method of measuring the defensive qualities of the bees we keep in Section 1.2 above and if our method is used it is highly unlikely that any nuisance is likely to be caused by stinging. To re-cap:

Followers up to say 2 to 3 metres	OK for a small garden apiary.
Followers up to about 5 metres	OK for an out apiary.
Followers up to about 10 metres	Unsuitable for any apiary and re-queening is required.

We consider that the use of gloves should be dispensed with except for dire emergencies. If the bees cannot be handled without gloves, then there is something wrong with either the bees or the beekeeper, both of which can be corrected. We record the number of stings we receive from each colony on the record card and this is also a numerate way of measuring the defensive qualities of the stocks. We don't expect to get stung during an inspection but if we do then there must be a reason for it and we do our best to assess the situation. If we wore gloves as the norm we would be unlikely to assess the temper of a colony in this way.

Finally on this point, if the colony is becoming difficult to handle then close it down; contrary to what many experienced beekeepers do, it is very good beekeeping practice.

1.7.2 Swarming.

The second most common form of nuisance to neighbours is the issue of swarms particularly if they fly over the neighbouring property. Most people that do not keep or have never kept bees suffer from api-phobia; that is, they have a great fear, aversion or hatred of bees. While it may be difficult for the beekeeper to comprehend this affliction, nevertheless (s)he must respect it. Therefore, if bees are being kept in a home apiary with adjacent neighbours, then swarming must be controlled ie. swarms do not issue from any hive in the garden apiary.

This involves three distinct husbandry operations, namely, swarm prevention, swarm detection and finally swarm control. These operations are fully discussed in Section 2.11.

1.7.3 Fouling.

The third and final most common nuisance is the fouling of washing and cars as a result of the honeybee defaecating on the wing. This aspect must be addressed at the planning stage of a home apiary in order to decide the probable flight path of the bees from the hives in the apiary. It is very easy to note where the neighbours hang out their washing to dry and to site the apiary where the bees are unlikely to over fly that particular area. The same reasoning goes for the parked car which quite often is its owner's pride and joy.

We have this problem at our own home apiary, not with washing or cars but with our boat which is moored at the bottom of our garden. She is our pride and joy. It often happens that she is fouled by our own bees particularly in the spring after her annual refit and all the topsides are polished and gleaming!

1.8 The Candidate will discuss methods of beekeeping and how these have been influenced by local conditions. Reference will be made to the choice of hives from those types commonly in use in the area, the merits of top and bottom bee space and apiary equipment.

1.8.1 We have difficulty understanding what is required of the Candidate when s(he) is asked to discuss methods of beekeeping and how they have been influenced by local conditions. We believe that the methods may be divided into two major headings each containing a variety of methods as shown below:

  a) Leave alone beekeeping
        Using skeps or log hives
        Using moveable frame hives

This category requires the beekeeper to maintain a daily watch for swarms and precludes the monitoring of disease or infestation conditions. We don't think that this is what the BBKA Examination Board is driving at in this part of the syllabus. However, the Candidate should be aware that there are many beekeepers in the UK who do just this.

  b) Regular colony inspections using moveable frame hives
        In fixed apiaries
        For migratory beekeeping (honey cropping and pollination)

As a hobbyist
As side-liner
As a commercial beekeeper

This category we believe contains the subject matter that the Candidate is likely to be questioned on.

The syllabus then goes on to enquire how these have been influenced (past tense) by local conditions, for example, the destruction of miles of hedgerows by the farming community to provide very large fields to allow cereals to be farmed as economically as possible. This has destroyed much valuable bee forage as well as destroying large areas of habitat for other wild life in some parts of the country. However, we would have thought the syllabus should read how they are influenced (present tense) by local conditions. In this case it would take in the local countryside as it now is, the geographical position and the crops available for honey and pollination. For example, the west side of the UK is generally wetter than the east side and crops are generally two weeks earlier in the south compared with the north.

A typical season for both honey and pollination would start with plums followed by pears, cherries, apples, rape, sycamore, horse chestnut, strawberries, raspberries, beans, lupins, mustard, white clover, bramble and finally heather to name some of the major sources. The flowering period of some of the above will overlap and not all hives will be working the same crop at a particular time.

It is essential, in our opinion, that the Candidate is familiar with the forage in the area around the apiary where the assessment takes place noting that there can be a marked difference between apiaries which are only a short distance (½ to 1 mile) apart.

1.8.2 Hives commonly used in the areas of UK.

1.8.2.1 General.

Frame and hive size. The size of the frame and the brood chamber are irrevocably connected one with the other. In general, the size is dimensioned by the fecundity of the queen and strain of bee aiming to operate with a single brood chamber with 10 to 12 frames. The British Standard (BS) frame and national hive were based on the performance of the Old English Bee which at best could produce only 8 frames of brood leaving 2 or 3 for brood chamber stores during the season. Such low fecundity does not exist with present day bees whether they be mongrels or of a purer line. The demise of the native bee occurred about 1920 when new strains were introduced into UK; it is therefore surprising that hives using the BS frame have continued to be so popular in this country.

Double walled hives. The principle involved here is that the two walls provide better weather protection which cannot be denied. It increases costs which also cannot be denied and they are unsuitable for migratory beekeeping. However it should be noted that one or two commercial beekeepers have used them for migratory purposes in the past; no doubt when labour was relatively cheap.

Single walled hives. The principle of the single walled hive is its constructional simplicity compared with the double walled types. All except the Modified National are of very simple construction involving only four pieces of wood to form the sides and all utilise frames with short lugs to allow this simplicity of construction. In the case of the Modified National extra timbers (4 in all) have to be used to form the rebate. Being simple makes them more suitable for migratory beekeeping.

Square versus oblong horizontal section. The square section is said to facilitate stacking hives when moving them from one crop to another on lorries or other vehicles. We doubt the validity of this and cannot think that a square could be more advantageous than a rectangle.

Top or bottom bee space. The principle is the same no matter whether it above or below - personal preference is a major factor when selection is made. See Section 1.8.3.

Long hives. One of the principles with this hive is that it not necessary to lift heavy supers at difficult heights and its low profile keeps it out of sight (neighbours and vandals) and sustains less wind pressure in winter gales.

Measurements. As all the hives discussed in this section were designed on the Imperial System of measurements (feet, inches and fractions of an inch), we have used Imperial measure first followed by the metric (millimetres) measure where appropriate. It should be noted that the conversions from Imperial to Metric are to the nearest mm.

It is unlikely that any list of hives will be complete as there will always be someone somewhere using either a hive of very old design or one which is not considered popular. However, the list should include the following:

Those using BS frames:

The Modified National (probably the most widely used).
WBC (with plinthless outside lifts).
The Smith Hive (frames with short lugs still popular in Scotland).

Those using other frame sizes:

The Modified Commercial.
The Langstroth.
The Dadant.

We believe that it essential for anyone contemplating this module of the examinations to actually see each of the above hives, take them to pieces, and make their own notes about them. Below we give a brief description of each and the BS frame, emphasising the important points only.

1.8.2.2 **The Modified National hive.**

This developed from the old British National which had finger tip lifting recesses, two single walls and two double walls which itself had developed from the Simplicity hive (1920). The distinctive rebates on the ends of the hive provide the accommodation for the long lugs and also make it into a truly single walled hive which is very easy to lift, the rebates providing a good hand hold. It was incorporated into BS1300 in 1946 and revised in 1960 but which has since lapsed. The salient points are as follows:

- Outside dimensions $18^{1}/_{8}$in x $18^{1}/_{8}$in x $8^{7}/_{8}$in (460mm × 460mm × 225mm)
- Inside dimensions $14^{5}/_{8}$in x $16^{5}/_{8}$in (371mm × 422mm)
- Normally constructed with bottom bee space but can be made with top bee space if required.
- Supers are constructed in an identical manner but with a depth of $5^{13}/_{16}$in (148mm)
- The floorboard 2in (51mm) deep with the actual floor $^{7}/_{8}$in (22mm) from the top edge to allow an entrance block $^{7}/_{8}$in x $^{7}/_{8}$in x $16^{5}/_{8}$in (22mm × 22mm × 422mm)

1.8.2.3 **The WBC hive.**

The forerunner of this hive was the Woodbury hive in 1860, the first double walled hive in Britain. William Broughton Carr designed his hive in 1890 which had lifts with parallel sides and plinths to locate one on top of the other. In 1899 James Lee & Son modified the design to the one we know today with sloping sides to the lifts and without locating plinths. The design has remained unchanged to this day and is still to be seen in the appliance manufacturers' catalogues. The salient features are as follows:

- The floorboard is of a special sloping design and stands on four legs splayed out at each corner.
- Slides are incorporated in the first lift for closing the entrance to the hive beneath an entrance canopy.
- All other lifts are identical and symmetrical.
- The roof is gabled with two sloping areas and the front and rear gables were traditionally fitted with a conical bee escape.
- The brood chamber and supers inside are constructed of very light weight timbers and each holds ten BS frames. There is no room for a dummy board in these boxes.
- Generally more expensive than the single walled equivalent.

We have not given the outer dimensions of the inner boxes as they tend to vary slightly from maker to maker depending on the weight of timber used. The inside dimensions are 15in x $16^{5}/_{8}$in x $8^{7}/_{8}$in (381mm × 422 × 225mm).

1.8.2.4 **The Smith hive.**

This hive was designed by W. Smith in Scotland based on a single walled hive used in America and with top bee space using BS frames with shortened lugs ¾in (19mm). The construction is similar to a British National and again has finger tip lifting recesses. It is constructed with timber of thickness $^7/_8$in (22mm). The salient features are as follows:

- Outside dimensions $18^1/_4$in x $16^3/_8$in x $8^7/_8$in (464mm × 416mm × 225mm).
- Inside dimensions $16^1/_2$in x $15^5/_8$in (419mm × 397mm).
- Supers are constructed in the same way but have a depth of $5^{13}/_{16}$in (148mm) compared with $8\ ^7/_8$in (225mm) for the brood box.
- Floors, crown boards, roofs, etc. are all dimensioned to fit on a similar basis to the Modified National.
- The hive is still very popular in Scotland.

1.8.2.5 The Modified Commercial hive.

The hive was designed by Simmins (date unknown) but mention is made of it in his book 'The Modern Bee Farm' published in 1887 so the hive must have existed at a date earlier than the date of publication of his book. Simmins held strong views about the inadequacy of the size of the BS frame and the hives designed around it. The Commercial hive, sometimes called British Commercial and more often Modified Commercial, uses Simmins frames 16in x 10in in size. We have not been able to trace whether the original hive was modified at any time as its name would suggest. Like the Smith hive it uses finger tip lifting recesses and it is not easy lifting a well filled brood chamber of this size by this means. The Commercial frames have short lugs in order to allow the hive to be square in cross section. The salient features are as follows:

- Outside dimensions $18^5/_{16}$in x $18^5/_{16}$in (465mm × 465mm).
- Inside dimensions $16^9/_{16}$in x $17^5/_{16}$in (421mm × 440mm).
- Supers are constructed the same way with a depth of $6\ ^3/_8$in (162mm) compared with $10^1/_2$in (267mm) for the brood chamber.
- As the outside dimensions are only $^3/_{16}$in (5mm) greater than the British National hive it is possible to use National floorboards, supers, roofs, crown boards and queen excluders with a Commercial brood box.
- Normally constructed for bottom bee space use.

There is only one point about this hive which is detrimental and that is the continual use of the bent metal runners which were designed for the National hive where long lugs on the frames are the order of the day. As the rebate in the Commercial hive is only just large enough to accommodate the runners ($^3/_8$in wide and $^{11}/_{16}$in deep) and they taper down into the corner of the recess, the taper being less than a bee space, all strains of bee are encouraged to propolise this area making manipulation difficult. The problem is easily solved by using runners of flat strip metal $^1/_{16}$in thick, $16^9/_{16}$in long by 1in deep attached with number 6 stainless steel pan head screws which then leaves a $^3/_8$in slot behind each runner which is the maximum dimension for a bee space. This space is not propolised. This is a good example of the equipment suppliers continuing to provide unsatisfactory equipment.

1.8.2.6 The Langstroth hive.

13

This hive must be the most used hive in the world and predominates in the major honey producing countries such as USA, Australia and New Zealand. In this country it is not used widely. The hive is a single walled hive with either top or bottom bee space using a frame $17^5/_8$in x $9^1/_8$in (448mm × 232mm) of comparative area to the Commercial frame. It can be designed to house either 10 or 11 frames. The salient features are as follows:

- Outside dimensions 20in x $16^1/_4$in (508mm × 413mm).
- Inside dimensions $19^1/_8$in x $14^1/_2$in (486mm × 368mm).
- Supers made the same way with a depth of $5^3/_4$in (146mm) compared with $9^7/_{16}$in (243mm) for the brood chamber.

There are variations of this hive perhaps the best known is the Langstroth Jumbo which has a depth of $11^3/_4$in (298mm) and uses Modified Dadant frames which have the same length as the normal Langstroth frame. Sparsholt College used them since the 1960s with crown boards with 4 ventilation holes $1^1/_8$in (29mm) diameter 3in (76mm) in from each corner.

1.8.2.7 The Modified Dadant hive.

The modification of this hive from the original one designed by Charles Dadant is well documented and is a mixture of ideas from Langstroth, length of frame $17^5/_8$in (448mm) and Quinby, depth of frame $11^1/_4$in (286mm). It is single walled and uses top bee space. The salient features are:

- Outside dimensions 20in x $18^1/_2$in (508mm × 470mm).
- Inside dimensions $19^1/_8$in x $16^3/_4$in (486mm × 368mm).
- Depth $11^3/_4$in (298mm) and $6^5/_8$in (168mm) for brood chamber and super respectively.
- It is capable of holding 12 frames with $1^3/_8$in (35mm) spacing or more usually 11 frames and a dummy board.

The best known variation of this hive is the Buckfast Dadant designed by Bro.Adam which is square in cross section and can accommodate 12 frames plus a dummy board.

1.8.2.8 Other hives.

There are many other hives being used in the UK and for completeness we include a brief description of some of them:

The Catenary hive. The basis is the catenary curve (the shape of the bottom edge of the comb built naturally by the bees in the wild). It was designed by Bill Bielby in 1968 when he was the CBI for Yorkshire based in Wakefield. It was found that top bars with wax starters were all that was necessary for the bees to build comb which they did not attach to the side of the brood chamber. National supers were used and the entrance was through a disc with the frames orientated the 'warm way'.

The British Deep hive. This hive is identical with the Modified National but $3^1/_2$in (89mm) deeper and takes a frame called a British Deep, similar to a BS frame but $3^1/_2$in (89mm) deeper. Appliance manufacturers make the frames and the hives. National supers are used.

The Dartington Long hive. This hive was designed by Robin Dartington in the early 1980s and is based on the British Deep frame. The length of the hive is such that it can accommodate 21 frames plus dummy boards and queen excluders. Provision is made for supers on top if required but normally honey is stored to the sides of the brood nest.

The Burgess Perfection hive. Not now manufactured but was made originally by Burgess of Exeter at the beginning of the 20th century. It is a double walled hive with the outer lifts fitting one inside the other for wintering and then reversing them to provide space for supers in the summer. We have seen some still in use in Devon.

In addition to the various hives for full colonies, there are also many kinds of different nucleus boxes that have been designed for queen rearing and mating using full frames, half frames, mini and micro frames. Ingenious ideas have developed for uniting adjacent nuclei and methods of overwintering them have been evolved. They are too numerous to discuss here and are outside the scope of the syllabus.

1.8.2.9 A description and measurements of various types of frame used in the UK.

It is important to thoroughly understand the concept of bee space before considering the pros and cons of various frames available commercially for use in different types of hive. Bee space is addressed in the Basic Examination (see section 2.2 in Beekeeping Study Notes for the BBKA Basic Examination - the Brown Book) and promptly ignored in all the other syllabi. We regard it as so important that any candidate for this examination should refresh their understanding of this very basic beekeeping concept. Nominal bee space = $^5/_{16}$in (8mm).

It should be noted that as all initial work undertaken in UK and USA on frames and bee space was done in imperial units, therefore, we have given them preference.

To understand the various types of frame available it is vital to actually see the different types and preferably to handle them; it is an unbelievable jungle and difficult to understand why so many variations are still available for purchase, particularly when many have so little to offer in the way of advantages.

1.8.2.9.1 Frames general: they have a top bar, two side bars and two bottom bars. They can have long or short lugs to suit the hive type and they are jointed to allow assembly without glue using only thin nails (gimp pins). They are better assembled using glue and copper pins giving a more robust construction with an infinitely longer life. There are two types of top bar: wedge type and slotted with a saw cut. Both are for fixing wax foundation onto the top bar without recourse to melting wax. The slotted type is a haven for the wax moth to pupate in. Side bars come in two types; parallel sided or with shoulders to provide self-spacing (eg. Hoffman), both types usually have a shallow slot as a guide for wax foundation on the inner surface. Bottom bars also come in two types; wide and narrow, the wide type being more robust and discourages the building of brace comb below the frame. All types of frame use wood $^3/_8$in (10mm) thick for the top and side bars.

1.8.2.9.2 Brood frames: top bars are available in two widths namely $1^1/_{16}$in (27mm) or $^7/_8$in (22mm) and similarly the side bars; it is possible to mix these with wide top bars and narrow side bars. Spacing of brood frames can be either $1^3/_8$in (35mm) or $1^1/_2$in (38mm) nominal, $1^9/_{20}$in (37mm) = the width of a metal end. When metal or plastic ends are used then the ends of the top bar must be $^7/_8$in (22mm) wide. Self-spacing is generally achieved with Hoffman frames which gives a space of $1^3/_8$in (35mm) between the centre line of each frame. It should be noted that a feral colony building its own comb from scratch, will have a spacing of $1^3/_8$in (35mm) between the centre line of each adjacent comb. If the bees build this way when left to their own devices, one may ask, why use spacers of $1^9/_{20}$in (37mm) and end up with $1^1/_2$in (38mm), give or take a bit of propolis.

Consider now the spaces between top bars and side bars when the two brood chamber spacings are used with the two top and side bar widths.

> Using $1^3/_8$in (35mm) spacing:
> > with $1^1/_{16}$in (27mm) frames, the distance between adjacent frames = $^5/_{16}$in (8mm)
> > with $^7/_8$in (22mm) frames, ... = $^1/_2$in (13mm)

> Using $1^1/_2$in (38mm) spacing:
> > with $1^1/_{16}$in (27mm) frames, the distance between adjacent frames = $^7/_{16}$in (11mm)
> > with $^7/_8$in (22mm) frames, ... = $^5/_8$in (16mm)

Only the first combination meets the criterion of bee space; i.e. between $^1/_4$in (6mm) and $^3/_8$in (10mm). Using this combination will minimise the building of brace comb between the woodwork of the frames.

1.8.2.9.3 Super frames: are usually spaced at $1^1/_2$in (38mm) and $1^7/_8$in (48mm) using metal or plastic ends or $1^5/_8$in (41mm) using Manley self-spacing frames. If the wide ends are used they can be staggered to reduce the spacing initially to $1^3/_8$in (35mm) if foundation has to be pulled out. Manley discovered (by experiment) that a spacing of $1^5/_8$in (41mm) was the maximum that can be tolerated for foundation to be drawn satisfactorily and produce thick combs of honey. The Manley frame has the further advantage of the top and two bottom bars being the same width, $1^1/_{16}$in (27mm), thereby providing a guide for uncapping and, of course, there are no ends to remove before extracting. Supers generally require a frame spacing wider than the brood chamber. During a honey flow 'wax builders' are in much evidence and, in the authors' experience, unless the bee space concept is observed much brace comb will be built in the supers. If $1^7/_8$in (48mm) spacing is used with $^7/_8$in (22mm) and $1^1/_{16}$in (27mm) wide top bars, then the distance between adjacent frames becomes 1in (25mm) and $^{13}/_{16}$in (21mm) respectively; very much larger than a bee space!

This simple arithmetic shows how little attention has been paid to the bee space concept in the design of frames and their spacing within the hive.

1.8.2.9.4 Other frames: are available but are not in general use, for example:

• made of plastic instead of wood. They are generally split in two halves in order to equip them with wax foundation.
• specially made frames for catenary hives. It was expensive to manufacture these to match the catenary of the brood chamber and most users just use a top bar; the bees do not attach comb to the walls of the hive catenary, a most interesting phenomenon.
• more recently an arrangement of specially manufactured plastic corners to accommodate the wooden bars without recourse to additional fixing nails or glue (trading under the name of 'Easy Beesy'). We have never tried them but in general we have an aversion to all plastic equipment for a variety of reasons.

1.8.2.10 Other types of spacers: are available but not very popular, for example:

• Plastic Hoffman adapters for converting frames with $^7/_8$in (22mm) side bars with metal or plastic spacers to self-spacing Hoffman.
• Yorkshire spacers which also attach to $^7/_8$in (22mm) side bars.
• Screws or studs in the side bars.
• Castellated spacers; usually 9, 10 or 11 slots for say a National and should only be used in supers, never in the brood chamber They save removing metal ends before extracting.

1.8.2.11 The British Standard frame.

As this frame is the most widely used in the UK, all Candidates for this module must be familiar with it including its dimensions. We believe that for examination purposes the dimensions of other frames need not be committed to memory but a general knowledge of the rough comb area is adequate with a knowledge of where to look up the exact details. It is wrong now to call this frame a British Standard because BS1300 has been allowed to lapse and has not been replaced. The salient features of the frame are as follows:

• The top bar can be either $^7/_8$in (22mm) or $1^1/_{16}$in (27mm) wide, the latter being preferable because with $1^3/_8$in (35mm) frame spacing, the distance between top bars then equals one bee space of $^5/_{16}$in (8mm) and reduces the brace comb building between adjacent top bars. Those top bars $1^1/_{16}$in (27mm) wide still have their lugs cut to a width of $^7/_8$in (22mm) in order to take a metal or plastic end spacer. Practically all top bars have wedges for fixing foundation (slotted top bars are not in much evidence these days).

• The side bars are normally $^7/_8$in (22mm) wide but can include the Hoffman shoulders to provide self-spacing without recourse to additional spacers. All side bars are slotted on the inside to provide a guide for the wax foundation.

- The two bottom bars are $^7/_{32}$in (5 or 6mm) wide and when fitted provide a slot $^1/_8$in (3mm) wide for inserting the wax foundation This slot is inadequate and makes it difficult to re-wax the frame without removing one of the bottom bars.

- The thickness of timber throughout is $^3/_8$in (10mm) except for the bottom bars which are $^1/_4$in (6mm).

- The overall size is 14in x $8^1/_2$in (356mm × 216mm) with a 17in (432mm) top bar giving lugs of $1^1/_2$in (38mm) at each end. The comb area of one side = 104 sq.in (67097 sq.mm) approx.

We believe that the design of the frame could be radically improved particularly in respect of the bottom bars. Our own design uses bottom bars $^5/_{16}$in (8mm) square thereby allowing a slot of $^1/_4$in (6mm) making the frames more robust and, easier to assemble and also easier to re-wax later. Additionally the design can be improved by replacing the Hoffman knife edge with a simple round or pan head screw in order to minimise the build up of propolis.

1.8.3 The merits of top and bottom bee space and apiary equipment.

With bottom bee space it is necessary to have a bee space below each box of frames and below each piece of associated equipment such as queen excluders, feeders, crown boards, etc. This makes the construction of the associated equipment more complex compared with top bee space as no surrounding plinth is required thus making for a simpler and cheaper form of construction. We have yet to find a commercially made piece of associated equipment using bottom bee space that matches the varying adjacent wall thickness of hives such as the Modified National.

A decision has to be made whether to opt for top or bottom bee space which must be observed throughout the hive and throughout the hives in the apiary to allow interchangeability one with another. The most common arrangement in the UK is bottom bee space which is curious when it is argued that with top bee space there is less likelihood of squashing bees during a manipulation compared with top bee space.

The major disadvantage, in our experience, with top bee space suggests that it makes the removal of a frame with short lugs more difficult during a manipulation; this, however, can be attributable to poor frame design encouraging propolising the frames to the hive body. The other disadvantage of top bee space seems to be that because the most popular construction in the UK is bottom bee space, the resale value of second hand equipment built with top bee space is lower.

After many years of beekeeping, if we were starting off from scratch again, we would opt for top bee space. The matter must remain one of personal choice.

1.9 The Candidate will describe the procedures for general maintenance including

preservation of hives, fumigation of comb and equipment, prevention of wax moth damage, the use of predator guards, storing combs and general apiary hygiene.

1.9.1 Procedures for general maintenance including preservation of hives.

There are no agreed procedures laid down, to the best of our knowledge, for the general maintenance and preservation of bee hives. The amount of maintenance required will depend on how well the hive has been constructed initially. Therefore there are three variables to consider, construction, preventive maintenance and corrective maintenance. Each will be examined in the following paragraphs.

1.9.1.1 Construction.

The following notes are concerned primarily with the construction of hives in wood, the traditional building material for bee hives.

1. A decision has to be made whether to opt for top or bottom bee space which must be observed throughout the hive and throughout the hives in the apiary to allow interchangeability one with another. The most common arrangement in the UK is bottom bee space which is curious when it is argued that with top bee space there is less likelihood of squashing bees during a manipulation compared with top bee space.

2. Because of the interchangeability requirement, construction must be to a high standard with manufacturing tolerances of not greater than $\pm^1/_{32}$in (1mm). Very few manufacturers achieve this standard.

3. For long life and minimum maintenance all joints in wooden hives should be dovetailed, glued and fastened with non-ferrous nails and screws. Failure to observe this will lead to nail corrosion and subsequent rot of the woodwork. Nail sickness, as it is called, is very common and can be seen in most apiaries.

4. Material for hive building is traditionally in wood which should be well seasoned (less than 15% water content). Red cedar is the lightest timber in weight with durable qualities but it is the most expensive in this day and age. Soft woods (deal and pine) are cheaper and quite satisfactory giving a long life but they require more maintenance than cedar. The main drawback is the proliferation of knots in soft wood. To prevent rot in soft timber it is very often pressure treated with a fungicide such as 'tanalith'. This is a trade name for an arsenic based fungicide which is toxic to bees and care should be taken to avoid it for construction purposes. In our opinion creosote is the most cost effective preservative for timber.

5. It is important that the construction is square and that there is no skew, a very common fault in home assembled equipment which can easily be obviated by ensuring that all construction work is done on a large flat surface.

6. The runners in supers and brood chambers are another critical part of the construction. When fitted they must provide a bee space below the lugs of the top bars. Most metal runners from the manufacturers require adjustment to obtain the correct $^5/_{16}$in (8mm). Metal runners are really the only acceptable material because of the requirement for disinfection by scorching with a blow lamp. Plastic runners are now appearing in order to cut down on costs; false economy in our opinion.

19

7. Roofs are very often provided by certain manufacturers too large for the hive and thus are easily blown off in a gale of wind in winter. The clearance between the inside of the roof and the outside of the boxes should be about $^5/_{16}$in (8mm). All corners should be dovetailed, glued and fastened with non-ferrous nails. Additionally, copper strengthening plates should be fitted at each corner on the lower edges to prevent distortion when under load with heavy supers during a summer inspection. Manufacturers used to fit these years ago but again they have disappeared as a cost cutting exercise. All our own hive roofs have them but they have been added to standard equipment. Metal covering for the roofs in galvanised steel is essential and they should have a layer of sheet polystyrene on the underside of the metal to minimise the temperature at the top of the hive in summer. The metal should be fixed with screws and not galvanised tacks. The roof ventilators require metal gauze (about 12 mesh per inch) on both the inside and the outside to prevent silent robbing. We have not yet come across a manufacturer who does this and most provide a mesh far too small thus restricting a ready flow of air. Most roofs have an inadequate spacer around the inside thereby not providing a large enough volume above the crown board. We keep our entrance block/mouse guard in the roof space of the hive and if the space is inadequate it will not fit in. All of our roofs have been modified because of this requirement.

8. Floorboards should be made of a weatherproof plywood rather than the old fashioned planked type to obviate cracks and joints in the floor where debris collects.

The main features in the construction of wooden hives have been described. Wood is the accepted material which, of course, was selected by the bees themselves many millions of years ago. Other materials have been tried for hives as follows:

Concrete reinforced with chicken wire (rather on the same construction principle as concrete boat hulls) for use in termite infested areas, eg. Africa. Termites can very quickly destroy a wooden hive and the log hive in Africa is normally hoisted by a rope into the branches of a tree to avoid termite damage. It is lowered by the same rope when it is required to inspect it or collect the honey crop.

High Density Polystyrene. This has been tried on account of its low thermal conductivity providing better insulation than most other materials. Impossible to disinfect with a blow lamp!

Other plastics. Many other types of plastic have been tried such as synthetic resins (reinforced with fibre glass) and polymeric substances. Most of these suffer from condensation problems and have insufficient porosity.

Metals. In the West Indies and Africa hives have been made from old oil drums.

No material seems to have surpassed wood for general use and cedar was the preferred timber as it was durable and light in weight compared with other hard woods of comparable durability. It is expensive these days and the soft woods and laminated shuttering are now in vogue. If laminates are used it is essential to ensure that the glues used for laminating are waterproof (take a test piece and boil it for half an hour in water - if delamination occurs then don't use it).

It goes with out saying that bee space is observed in the construction of all the hives discussed. We regard bee space to be $^5/_{16}$in (8mm) to allow a tolerance of plus or minus $^1/_{16}$in (2mm) for expansion or contraction when the hive is in use. This means that the bee space can vary between $^1/_4$in (6mm) and $^3/_8$in (10mm) which is within the limits for use by the bee without propolising or building brace comb.

It will be clear from the above that there is a lot wrong with hive equipment from the bee appliance manufacturers which can only be corrected by the beekeeper demanding higher standards.
The following notes are concerned primarily with making frames using wood, the traditional material for their construction.

The main features in the construction of (perhaps assembly for most beekeepers) frames may be summarised as follows:

1. It is important that each frame be assembled so that it is square and free from skew otherwise it will upset the bee space when the frames are used in the hive either one frame to its neighbour or the frame with respect to the side walls of its box.
2. For long life all joints must be glued (eg. Cascamite or similar waterproof glue) and fastened with copper nails. Copper nails do tend to be rather thick and pre-drilling is advisable to prevent the timber from splitting. Compare frames a few years old assembled without glue and fastened with steel gimp pins with those assembled as described. The joints will be slack and there will be evidence of nail sickness in the surrounding wood. An old frame should be quite rigid before it is re-waxed with new foundation. A reasonable pressure on opposite corners trying to force the frame into a parallelogram will be a satisfactory test.
3. In general the nails fastening the joints should be at right angles to the direction the joint is put together. The wedge top bar fixing the foundation should be pinned horizontally; so many times one sees frames where the pins have been put in vertically from below and the points protruding from the upper surface of the top bar precluding scraping with a hive tool.

Other points of interest.

1. We consider that there is a case to be made for the re-design of the frames that are used in the hives used in the UK as most of those available have design faults.
2. The following leaflets are worth seeking out and studying:

a) BBKA Advisory Leaflet "The preservation of beehives and their ancillary equipment".
b) MAFF(now DEFRA) Leaflets which are exceptionally good but unfortunately all are out of date and unlikely to be reprinted. Many of the older beekeepers have copies and some may be found in association libraries. The following should be sought:
Number 144 - Beehives
Number 367 - The British National Hive 1979/1982
Number 445 - The Smith Hive 1960/1981
Number 468 - The Modified Commercial Hive 1982
Number 549 - The Langstroth and Modified Dadant (MD) Hives 1980

3. 'The Illustrated Encyclopedia of Beekeeping' by Morse and Hooper has good sections on hives and frames together with good photographs. 'A Case of Hives' by Heath is also worth reading for this part of the syllabus.

1.9.1.2 Preventive maintenance.

Perhaps it may be constructive to start by defining preventive and corrective maintenance which is applicable to any type of equipment. Preventive maintenance is those procedures/methods/actions deliberately taken to prevent a breakdown when the equipment is in normal use. Corrective maintenance is those actions required to repair the equipment in order to bring it once again to a serviceable state after it has broken down.

There are a few preventive measures to be taken as follows:

a) All wooden hives should be placed on suitable stands to keep the woodwork away from the ground, preferably on concrete or on a material impervious to water to minimise rot.
b) All the woodwork both inside and outside should be disinfected by scorching with a blow lamp to the colour of light straw to kill all the pathogens affecting the bees and those that cause wet and dry rot in timber.
c) All the woodwork outside the hive should be treated with a wood preservative again to prevent rot occurring.

We always maintain our brood chambers every two years, our floor boards and entrance blocks every year and our roofs every three years. Creosote is the cheapest wood preservative and very effective while other proprietary preservatives are no more effective, more expensive and a good way of getting rid of disposable income. We have some equipment that is over 50 years old and still very serviceable; it originally belonged to Taylors of Welwyn. For single wall hives paint should not be used as it does not allow the woodwork to 'breathe'.

1.9.1.3 Corrective maintenance.

The corrective measures that are required may be summarised as follows:

a) Scraping all equipment clean removing wax and propolis and polishing the metal runners with wire wool.
b) Disinfection by blow lamp to kill disease pathogens. It seems pointless to fumigate a brood box inside with acetic acid or similar. It would be difficult to treat the outside without a specially made fumigation chamber.
c) Any damaged woodwork will require to be repaired to ensure that everything is once again bee tight.
d) The runners will require to be greased.
e) Finally treat the woodwork outside with preservative again.

1.9.2 Fumigation of comb and equipment.

We have had some difficulty understanding the requirements of this part of the syllabus when it states that "the Candidate will describe the procedures for the fumigation of comb and equipment". Fumigation of comb and the supporting frame is straightforward but we are not aware that the fumigation has been applied to other items of equipment. Fumigation of comb can be undertaken with 80% acetic acid and paradichlorobenzene (PDB) but it should be noted that comb and equipment can be undertaken using gamma radiation but it is expensive and not a practical option for a part-time beekeeper. Additionally, fumigation using methyl bromide or ethylene dibromide, because of their toxicity, is not a practical option for the average beekeeper.

Therefore, in this section we will consider only the fumigation of comb which we believe is the intent of the syllabus. There are two reasons for fumigating comb:

a) To kill many of the pathogens found in brood comb with acetic acid.
b) To ensure that wax moth damage does not occur during storage.

Both these items are dealt with in detail in Section 1.9.3 and 1.9.5 which follow.

1.9.3 Prevention of wax moth damage.

1.9.3.1 Wax moth damage to stored comb.

1.9.3.1.1 Greater wax moth *(Galleria mellonella)*.

The adults have a wing span of 1 to $1^{1}/_{2}$ inches (25 to 38mm) and enter the hives at night to lay eggs. The eggs hatch to larvae and when fully grown they are about $^{7}/_{8}$ inch (22mm) long and quite distinctive with a dark head. The larvae pupate, usually in a boat-shaped groove chewed into the woodwork of a frame, the chrysalis eventually hatching to the adult form. The damage is caused during the development of the larval form. The larva has the ability to digest wax but it also needs protein which is obtained from pollen and larval debris of the honeybee.

The life cycle is approximately, egg - 7 days, larva - 15 days, pupa - 28/32 days. The times are very variable and depend on temperature (egg to adult on average is 50 to 54 days).

In warm weather there is the possibility of all the comb in a full brood chamber being turned to dust in about 14 days and much of the woodwork damaged if it is off the hive with no bees. A strong colony will not tolerate the moth and keeps itself in a healthy state and no damage is caused. Weak colonies can be damaged with the bees in occupation.

These moths are generally only troublesome in UK when the comb is not in use, they can however be very real pests in tropical climates.

It is possible to introduce bacteriological control by impregnating the stored comb with spores of *Bacillus thuringiensis* which kills the wax moth larvae. This form of control has been used in USA but is not practised widely in UK at the present time (It is sold under the trade name CERTAN).

1.9.3.1.2 Lesser wax moth *(Achroia grisella)*.

This moth which is much smaller and has a wing span of about $^3/_4$ to 1 inch (19 to 25mm) and weighs about a sixth to a tenth of the weight of the greater wax moth. The female generally lays eggs in crevices and it is generally accepted that the larvae do not cause damage to the woodwork. However, we have heard reports to the contrary but we, ourselves, have never experienced woodwork damage with these lesser wax moths. The larvae consume and digest the wax comb and while doing so they produce a large web of silk tunnels. It is not so much of a pest in this country as the greater wax moth but can completely ruin comb if an infestation occurs and no protective measures are taken.

1.9.3.1.3 Death's head hawk moth (*Acherontia atropos*).

This is a magnificent looking moth with a skull and cross bones marking on its thorax on the dorsal side. It is attracted to bee hives and is quite rare in UK; it originates from N. Africa and Spain. It is worth having a look at one in a natural history museum. It has been reported in Devon that more frequent sightings of these moths have been seen during 1998. Whether they are on the increase or whether they are taking advantage of weaker colonies during a poor honey season remains to be established.
The adults feed mainly on sap from tree wounds but they can resort to robbing nectar and honey from the hives of honeybees usually attacking the hive at night. They are not generally a pest in this country as can be noted by the *Acherontia* cadavers found occasionally in the hives of the honeybees.

1.9.3.2 Methods of storing comb with particular reference to prevention of wax moth damage and sterilisation against nosema.

1.9.3.2.1 Types of comb to be stored: are supers and brood comb.

a) Supers. These can be stored either wet (with honey) or dry after being cleaned up by the bees and removed from the hive again. This comb consists only of wax and honey (if stored wet).
b) Brood combs. This type of comb is very different containing wax, pollen, larval skins and faeces, propolis, etc. making them much more attractive to attack by other insects and mammals.

1.9.3.2.2 The main causes of damaged comb are by;

a) Mammals such as mice, rats, squirrels, etc. which are easily excluded with travelling screens or queen excluders at the top and bottom of the stacks of boxes of frames.
b) Insects, the main cause of damage being the wax moths. There are two:
   1. The lesser wax moth (*Achroia grisella*) and
   2. The greater wax moth (*Galleria mellonella*) which is regarded as the major pest.
However both can cause very extensive damage in a short time if precautions are not taken.

1.9.3.3 Methods of protection against wax moth damage.

There are four methods namely paradichlorobenzene (PDB), acetic acid (80%), heating and cooling. Each of these either kill (k) or have no effect (ne) on the various stages of development ie. egg, larva, chrysalis and adult. Reference to published literature reveals the following:

	PDB	ACETIC	FREEZING	HEATING
Eggs	ne	k	k	k
Larva	k	ne	k	k
Chrysalis	?	?	k	k
Adult	k	k	k	k

T (FREEZING) = 0°C to -17°C for a few hours to a few days depending on temperature and bulk of frames.

T (HEATING) = 120°F (49°C). Note the melting point of wax ≈ 145°F (63°C)

? = no reference could be found in the standard literature.

It will be clear from the above table that the best method is freezing before storage at normal temperatures. No airing of the combs is necessary. A good method for supers where the risk of disease is very low compared with brood frames.

Heating would also be possible but wax is very malleable at 120°F (49°C) and the temperature control would have to be precise.

Gamma radiation is known to kill all stages in the life cycle but is expensive and not really a practical method for the beekeeper. For the average beekeeper fumigation is the more usual method and it will be clear that both acetic acid and PDB is necessary and is the accepted method of dealing with brood frames.

• Brood frames. These should first be fumigated with 80% acetic acid as follows:

a) Brood box with frames placed on suitable board (acetic acid attacks concrete).
b) Metal runners in brood box well greased to protect from acid fumes. Any metal ends are removed from the frames.
c) An empty eke is placed over the brood box and covered by a suitable board.
d) The acetic acid is poured onto an old piece of rag in a shallow dish standing on the tops of the frames. The rag extends over the edge of the dish acting as a wick. The fumes are heavier than air and fall through the frames. The amount of acetic acid required is 100ml per BS brood box with 11 frames or pro rata for other sizes.
e) All the joints should be sealed with tape to make the set up as air tight as possible.
f) Leave for one week and then the frames can be stacked for winter storage.
h) All the exposed wooden parts of the frames should be scraped clean of wax and debris before fumigation.

After fumigation with acetic acid they should be stacked as follows:

a) Mouse excluder with newspaper over.

b) Sprinkle one dessertspoonful of PDB crystals onto newspaper and place brood box over.
c) Cover with newspaper, PDB, another brood box, etc. finishing with a screen and crown board.
d) The PDB should not come in contact with the wax comb where it will contaminate the wax.

The above treatment will, before storing, kill any pollen mites (*Carpoglyphus lactis*) a harmless minute mite that burrows into pollen filled cells scattering the contents over the combs. It is the only way we know for cleaning out pollen clogged combs.

• Supers. It is unnecessary to fumigate with acetic acid and they can be stacked straight away with PDB and newspaper. At the end of the season it is important to get all frames cleaned up, fumigated and stacked for winter as soon as possible. While the weather is warm the wax moth can do considerable damage.

Note that if supers are stored wet it is necessary to make them bee-proof if they are stored outside otherwise they must be in a bee-proof shed or room. When the frames are wet, they are not attractive to wax moth so they can be stored without PDB. They do become very damp (honey hygroscopic) and tend to grow mould during the winter.

1.9.3.4 Other relevant points:

a) PDB crystals should never come into contact with the wax comb.
b) Disinfection with acetic acid is the approved method of cleaning comb infected with Nosema spores, so the storage treatment ensures that there is no risk of the spread of infection the next season when the comb is reused. Other pathogens are killed with acetic acid such as chalk brood fungus spores so it is good beekeeping practice always to disinfect brood combs before storage.
c) Rather than have the trouble of stacking and sealing boxes for acetic acid treatment it is probably better to have a permanent installation to hold as many frames as required. It can be custom built and made completely air tight thereby using less acetic acid. The disinfection box automatically can become a frame storage box in the winter.
d) After any fumigation, combs should be well aired before re-use in the hive.

1.9.4 The use of predator guards.

The major pests requiring consideration for successful wintering are mice, other mammal pests, birds and human beings. Each will be examined in the following paragraphs.

1.9.4.1 Mice. These include the common or domestic mouse and the field and wood mouse. They will enter hives in the autumn seeking somewhere dry and warm to build a nest for hibernation purposes. This activity is prompted by the shorter days and a drop in temperature. The moral is to have mouse guards on the hive in plenty of time to ensure that a mouse does not enter to feed and build a nest.

• Mice feed on pollen, honey and bees. They therefore cause damage to comb, frames and hive equipment. In winter they will disturb the winter cluster and this disturbance can kill the colony if the temperatures are very low. The bees can sting mice to death and they have been known to be embalmed in propolis because the bees cannot eject them from them hive. Any droppings and urine are generally cleared out by the bees.

• Mice have oval skulls and can squeeze through a $^3/_8$ inch (9mm) wide slot but they cannot pass through a $^3/_8$ inch (9mm) diameter hole. Mice are therefore not a problem to keep out of the hive and if they do enter, it is the fault of the beekeeper not taking the necessary precautions in time by providing suitable predator guards or mouse guards.

1.9.4.2 Other mammal pests. These include shrews, rats, moles, squirrels, hedgehogs, etc. All these can disturb an overwintering colony and in this respect can cause damage to it, but note that many of them are hibernating themselves. We have noticed pronounced scratch marks at the entrance to some of our hives at one apiary and believe it to be due to badgers although we have not caught them in the act.

1.9.4.3 Birds as pests. The main culprit is the green woodpecker in very cold weather. They usually peck through at the hand-hold on National and Commercial hives and in a matter of an hour can make a hole of sufficient size to enter. If they are not spotted in time the colony will surely perish for it will occur in periods of hard frost or snow on the ground. Combs, frames and hive parts will be damaged. Guards can be provided to prevent colony damage. See Section 1.9.4.6.

Other birds are swifts, tits, swallows, shrikes, etc. taking bees on the wing (including queens on mating flights). We have watched sparrows in the early morning sitting on top of the hive waiting for bees to come out, catching them and taking them back to their nest for the fledglings. Pheasants also have a taste for bees; we wondered why one colony at one apiary was very often irritable until one morning we saw a pheasant tapping at the entrance and eating the bees as they came out to investigate.

1.9.4.4 Good ventilation while excluding mice. A colony during winter, if it metabolises 35lb (16kg) of honey, will be required to get rid of approximately 4 gallons of water. This can only be achieved by evaporation. The average rate is 5 pints/month or 3 ozs/day. It is more difficult for evaporation to take place in the damp western side of UK compared with the drier eastern side. These are the facts, the best configuration for achieving this evaporation is still being debated in the bee press and still no one seems to agree on the subject.

Our own method, which we have used successfully for many years after experimenting with various approaches, is as follows:

a) Heat escaping from the cluster causes the movement of air, warm moist air moves upwards and is replaced by cold dry air at the bottom.

b) All entrance blocks have 9 x $^3/_8$ inch (9mm) diameter holes drilled in them spaced equidistant apart across the length of the block. This gives a total cross sectional area of c. 1 square inch (645mm^2) to limit the air flow. The block turned through 90° is a normal reduced entrance block with a 4" wide slot. The $^3/_8$ inch (9mm) diameter holes form the mouse guard and are 'kinder' to the pollen collectors in the spring compared with the perforated metal strips sold by the equipment suppliers.

c) The crown board is raised about $\frac{1}{8}$ inch (3mm) with matchsticks at each corner. This gives an exit area for air to escape of c. 9 square inches (5,800mm^2). The feed hole(s) are covered so that the flow of air is round the outside of the cluster and avoiding the chimney effect directly above the cluster. The roof ventilators now play no part in the ventilation system.

d) The mouse guards are put in usually in September before the ivy flow and the crown board is raised as late as possible to stop the gap being propolised. It is interesting to note that if there is no ventilation at the top of the colony, in the spring it is usually a mess with condensation and mouldy outside combs. One associated problem is that some of the stored pollen also develops mould and is then useless to the bees.

e) Strains of bee that produce a lot of propolis will get themselves into this situation if crown boards are raised too early.

1.9.4.5 It is very important to keep human beings out of the hive during winter or at the very least to stop them tinkering with it. Unfortunately there are no human guards available for fitting to a hive. No disturbance during the winter period is essential. Once the colony has settled down for winter it should be left undisturbed until the following spring. Experiments have been conducted and it has been found that the cluster temperature is raised quite a considerable amount (up to 10°F or 5.5°C) by say just taking the roof off. Such increases in temperature shorten the life of the winter bee and this manifests itself in spring just when the colony requires all the bees it can muster. Hives should never be sited under trees where the drip of water from the branches can cause colony disturbance.

1.9.4.6 Keeping the green woodpecker at bay.

The green woodpecker can spell disaster for a colony if it directs its attention to boring through the side of the hive. They are usually troublesome in very cold weather when they cannot find forage in the hard ground. There are two ways of protection:

- Surrounding the hive with chicken netting ensuring that it is kept about 6 inch (150mm) from the hive walls to prevent the woodpecker from reaching through to the woodwork from the netting which provides him with a good foothold.
- Covering the hive with a plastic bag (but note this interferes with the ventilation) and does produce condensation inside the bag. While it denies the woodpecker a foothold we do not recommend it because of ventilation and other problems.

1.9.5 Storing combs. (See Section 1.9.3)

1.9.6 General apiary hygiene.

This must be the easiest part of the syllabus and is a very popular examination question. Good apiary hygiene is summed up in the following , often referred to as the 10 commandments of good beekeeping:

1. Always keep the apiary clean and tidy.

2. Never throw propolis or brace comb on the ground; be sure always to place it in a suitable container and remove it from the apiary.

3. Never buy old combs.

4. Never buy colonies of bees unless it is known that they come from disease free apiaries; never accept stray swarms from unknown origins.

5. Always disinfect second hand hives and other equipment before use.

6. Never feed honey or allow bees to gain access to it; refined sugar is the only acceptable feed for honeybees.

7. If a colony dies out during the winter (or at any other time) and the trouble is not due to starvation, close the hive, pending the examination of a sample comb and bees, to prevent the remaining stores being robbed out.

8. Never exchange brood or super frames/combs between one colony and another unless it is known that all colonies are free from disease. Where possible, supers should be marked and always used on the same colonies.

9. Take care to prevent robbing at all times by observing item 2 and not spilling syrup or having leaky feeders.

10. Arrange all hives in such a way that drifting is reduced to a minimum.

The code is essentially for minimising disease and its spread in an apiary and appears in a modified form in ADAS Leaflet P306 - 'Foul brood of bees: recognition and control'.

There is only one other item that we would incorporate in the list and that is the requirement of renewing comb in the brood chamber once every three years on a rotational basis. It is another way of getting rid of disease pathogens. Having said that it is common practice in the USA to keep brood comb for year after year without apparent detriment.

1.10 The Candidate will describe the associated dangers of robbing and describe the methods in place to prevent robbing and to end robbing once started.

1.10.1 General points in relation to bees robbing bees.

• In nature, a concentration of colonies does not occur and therefore robbing is not a problem. It only occurs where the beekeeper has concentrated his stocks on to a single site to form an apiary. The beekeeper with only one stock will seldom have trouble with robbing.

• Robbers are generally bees from another colony but wasps, hornets and ants can also rob a hive. Ants are not a problem in the UK but overseas in the tropics they are a problem.

• Robbing is for honey only, the other hive products such as pollen and propolis attract no attention as plunder.

• Different strains of bees have different propensities to rob other colonies; the Italian yellow strains being the worst, they are inveterate robbers.

• It is more likely to start after a nectar flow has come to an abrupt halt and in times of dearth.

• It is usually started as a result of bad management practices on the part of the beekeeper.

• Robbing can occur between hives in a single apiary or between hives in two apiaries.

• When robbing occurs in an apiary the only method of communication between the bees is by the round dance which only gives information on distance. Because no directional information is available the bees can only search in the near vicinity which may initiate further robbing if a weak colony is discovered.

• It has been suggested, but not proven, that robbers may release a pheromone to mark the site to be robbed.

1.10.2 The dangers associated with robbing.

• When honeybees are involved there is a potential danger of spreading disease.
• Many valuable foragers are likely to die in the fighting which often occurs.
• Large amounts of honey are lost to the successful robbers, occasionally to another apiary and another beekeeper.
• When wasps are involved they very often kill off the colony because they are the stronger insect.
• There is a danger of stinging due to the ensuing hiatus.
• Weak colonies and nuclei are likely to be decimated.

1.10.3 Methods to avoid robbing.

• Prevention is always better than cure, and good apiary practice at all times is usually the answer.
• Because bees are only interested in a free supply of honey/nectar or sugar syrup available in quantity then there should be no spillage or traces of syrup outside any colony or within the apiary.
• There should be no way into any colony except via the designed entrance; all equipment should be bee tight.
• Colony entrances should be adjusted to the size (or strength) of the colony and to the time of the year and flow conditions.
• When there is no nectar flow, colonies should not be kept open for too long during manipulations.

1.10.4 Methods of detection of robbing.

There are two types of robbing. The first involves fighting at the entrance of the robbed hive and the second is called silent robbing where no fighting takes place at or within the robbed stock. The behaviour of the foraging bees is quite different in the two cases.

Silent robbing: is characterised by the robbed colony continuing to work normally while at the same time the robbers also enter and leave the robbed colony in a normal manner. The robbed colony can itself be robbing another colony at the same time. The only tell-tale sign is the flight of the bees returning directly to another colony in the same apiary.

One final last point on silent robbing. We discovered, quite by chance, many years ago the robber bees inside the robbed hive passing the spoils to their compatriots through the roof ventilators. Closer observation revealed this to be quite a common occurrence. The solution was simple. We have equipped all our hive roofs with protective wire gauze both inside and outside the hive making contact between the two parties impossible. It also has the advantage of preventing the roof ventilators becoming blocked by other insects.

Robbing with fighting: has two recognisable characteristics. The first is the fighting outside the robbed hive and the second is the flight of the robber bees approach which is nervous and erratic. The erratic zig-zag flight is curious because it alerts the guards of the robbed colony. Once the robber bee alights and is challenged it becomes submissive and often offers food to the guards.

The characteristic common to both types of robbing is the flight of the laden and unladen bee; rear legs forward in the first instance with a full honey sac and with rear legs trailing astern when unladen in the second instance. The normal rear leg position in flight is reversed, ie. a normal forager should not leave the hive with full honey sac and return empty.

1.10.5 Methods used to terminate robbing.

There is no effective way to stop robbing the day it starts. Removal of the robbed stock to another apiary is unsatisfactory as it usually gets robbed again at the new site (the colony being possibly marked by pheromone). The robbing stock is likely to find another weak stock and continue robbing. The following actions are all effective to some degree:

• Remove robbers to a remote site isolated from other colonies in the immediate vicinity.
• Reduce all entrances and make the nucs and weaker stocks a narrow tunnel (one bee space wide) about 2in (50mm) long.
• Straw and grass to cover the entrances of both the robbed and robbing hive to confuse both parties has been suggested by some writers.
• Plain glass leant up against the entrance allowing only entrance from the sides.
• Reversal of the robber and robbed colony.

If any signs of robbing do occur, we consider that the first action must be reduced entrances and this is why it is so important to have the hive entrance block always stored in the hive diagonally across the crown board when not in use. Nucs are particularly vulnerable and methods of restricting any nuc entrances immediately must be normal apiary management. Note that many of the equipment suppliers, economising on wood, do not make the trim in the roofs of hives deep enough to take an entrance block - they are very easily rectified by tacking 4 laths to the existing woodwork.

If a robbed colony is moved it is always wise to leave a frame with some stores in it on the site and allow the robbers to clean it out and finish the robbing job to their satisfaction (the one frame can be put in a spare nuc or travelling box).

1.10.6 Robbing by wasps.

Eleven species of wasp are found in Europe and seven of these are found in the UK. The most common are the *Vespula vulgaris* and then the *Vespula germanica* both of about the same size. Nests of *Vespula media,* a common continental species, have been found in the South West of England when the summers are long and warm. However, so far, there have been no reports of this species robbing beehives. Hornets, *Vespa crabro*, are also fairly common but in our experience seldom cause problems as robbers of bee hives.

The *Vespula vulgaris* is the main culprit and its robbing can be quite devastating if a colony exists nearby an apiary. This common wasp is physically stronger than the honeybee and in large numbers can easily overpower a quite strong colony. The nests are usually underground in old mouse holes or rabbit holes. As the colony grows so the nest size is increased and this often requires the wasps to enlarge the cavity by moistening the earth and removing it in small pellets to the outside while maintaining a small entrance. Their mandibles are also strong to do this work and to prepare the wood pulp which the nest is built with.

We know of no way of stopping wasps robbing once the process has started except by seeking out the nest and destroying it after dark when all their foragers (and robbers) have returned for the night. In 1995, the worst year we have known for wasps, we lost two colonies which were decimated by wasps.

It is not clear whether wasps communicate food sources to their nest mates as do the honeybees. However, the observation of colonies being killed off would indicate that they have some mode of communication considering the numbers involved in their concerted attacks. Examination of the combs of a colony killed by wasps will reveal all the honey stores have been taken together with the brood and only the exoskeletons of some of the worker bees remain, the soft internal viscera have been removed. Wasps are the best insecticide available!

1.11 The Candidate will describe the methods used to minimise drifting and circumstances when diverting bees to another colony can be an advantage.

1.11.1 The definition of drifting and general considerations.

It is generally accepted that drifting occurs when a bee leaves one hive and **mistakenly** joins another hive either as a result of confusion or as a result of the bee being blown off course by the wind.

When drifting occurs and the wrong worker bee attempts to enter the wrong hive, it is usually challenged by the guards. The drifting bee's behaviour is different to those of other worker bees entering the hive; it can readily be observed that it 'bribes' the guards by offering some nectar (they are usually returning with a full load) while it is being examined by them. No fighting occurs.

On the other hand, drones drift much more freely than worker bees and appear to be readily accepted into any hive generally without being challenged.

Drifting occurs most in young bees during the first 4 days of their adult life and N.E.Gary claims that this happens with about 20% of these young bees. This happens during 'play flights' when the young bees start to take orientation flights (also simultaneously cleansing flights). It is an unusual behaviour pattern because many young bees do this in quite large numbers and can be seen hovering in front of the entrance for a few minutes which ends quite abruptly and the activity at the entrance returns to normal. Subsequent play flights are extended in range away from the hive and this is when the drifting is more likely to occur.

Drifting only occurs in man made apiaries. It does not occur in feral colonies because they are so widely separated under normal circumstances.

1.11.2 Apiary configurations and associated disadvantages.

When hives (usually of the same colour and design) within an apiary are sited in the same direction in long rows severe drifting may be expected; and the closer they are together the more severe the problem. If not effected by a prevailing wind then it has been observed that the centre colonies of the row are weakened at the expense of the end colonies of the row. If there is a prevailing wind effect then the colonies to leeward will be strengthened.

If the apiary has more than one row of hives also facing in the same direction, then the front row will be strengthened at the expense of the rows behind (the bees drift forwards).

In the situations outlined above it has been noted, in America, that the difference in yield amounts to about 20lb per season in the hives that have been fortified with drifting bees.

The disadvantages of selecting a poor layout are as follows:

a) In large apiaries in rows it would be necessary to balance the colonies to maintain them at roughly equal strengths. When some colonies become very strong while nearby colonies are weak, then this is a recipe for robbing to start when a flow comes to an abrupt end.
b) If there is disease in a colony the bees drifting from this colony can carry the disease pathogens to the other colony. All the bee diseases can be transmitted to hives in the same apiary by this mechanism.

1.11.3 Methods of minimising drifting within the apiary.

The most important feature of any apiary layout to prevent drifting is to adopt an irregular pattern of hive layout with the hives facing in different directions and adopting curved lines, arcs, circles, etc. Perhaps the best regular pattern to adopt is having the hives in blocks of four each facing in different directions at right angles to each other (eg. N, S, E & W). These blocks of four can be repeated about 5 metres spacing between the blocks. We use this in one of our apiaries and find virtually no drifting occurs with worker bees.

Our home apiary of six colonies is by necessity in a long row (there is no other way) with 9 ft between each hive. Drifting is quite prevalent but is something that we have to accept. Matilda Herz showed by experiment that it was possible to train bees to recognise marks at their hive entrances. They are able to recognise the difference between solid and open patterns but unable to distinguish between solid shapes or between open shapes. We tried this in our own apiary by making the shapes in white formica about 6 in x 6 in and attaching them, alternately solid and open, onto the alighting boards of each hive in the row. We could not confirm the results claimed by Matilda Herz and drifting still occurred.

Other methods of providing aids to navigation within the apiary are:

   a) Hive entrances of different colours. This is very popular in Germany and elsewhere on the
   Continent.
   b) Providing a discrete marker at or adjacent to each hive entrance, eg. a stone, a bush etc.

1.11.4 When diverting bees to another colony can be an advantage.

In general and under normal circumstances, we consider that there is no clear cut advantage in
diverting bees to another colony. The only manipulations where we consider this acceptable is in
swarm control methods (for example, when making an artificial swarm whereby all the bees all
come from the parent colony and there is no fighting involved) and when colonies are brought
together for uniting. Another manipulation where diverting bad tempered bees away from the
parent colony (to a box of comb and not to another colony) is used when requeening a bad
tempered stock.

We have seen references from time to time about swapping round the position of a weak hive and
a strong hive in order to strengthen the weak colony or nucleus; we consider this to be nonsense.
If the weak colony needs strengthening then the proper way to do this is to provide it with a
frame(s) of emerging brood from a disease free colony. Then the beekeeper is in control and no
fighting will be involved.

1.11.5 Other points of interest.

• Queenless colonies are an attraction especially at queen rearing time. Although this is technically
not a drifting problem, virgin queens from swarms will drift in if in the near vicinity. We had our
queen rearing programme ruined one year by a stray swarm alighting on a tree in our home apiary
and virgins from the swarm entered our cell builder and destroyed all our cells!
• Our experience has been that no matter what apiary layout we have adopted, drones hop in and
out of any hives, willy-nilly, the whole season.

** ** ** **

PRACTICAL BEEKEEPING

The Candidate will be able to demonstrate understanding of, and ability to perform the following tasks:

2.1 Describe the role of good hygiene in the apiary. The candidate will demonstrate satisfactory procedures, both with personal effects and apiary equipment.

2.1.1 See Section 1.9.6 for details of the ten commandments for observing good apiary hygiene.

2.1.2 Demonstrating satisfactory procedures with both apiary equipment and personal effects.

We consider satisfactory hygiene for apiary equipment involves the following:

a) Annually - disinfect with blowlamp queen excluders and floor boards and then creosote outside surfaces of floorboards. Change three/four frames in each brood box using disinfected frames or frames with new foundation. Clean all super frames by scraping away surface wax and propolis; ditto super boxes. Clean and overhaul smoker inside and out. Clean up hive tool. We have never found it necessary to disinfect either the smoker or the hive tool unless they have been used on a colony infected with either of the foul broods.
b) Every two years - disinfect with blowlamp crown boards and brood boxes and then creosote the outside surfaces of the brood boxes.
c) Every three years - disinfect roof with blowlamp and treat outside surfaces with creosote. Treat outside of supers with creosote.

2.1.3 We are at a loss understanding what the Examination Board were requiring in respect of 'satisfactory procedures with personal effects'. We have scoured the classical literature and have not been able to find anything of any real moment on the subject. The only personal effects associated with beekeeping and the apiary must be the bee suit or jacket/smock, veil (which may be incorporated in the bee suit but is removable), gloves and wellies. Clearly all these would be disinfected if there had been any association with the foul broods, but under normal circumstances they would be washed as required. Don't forget to get them to the dhobi prior to the examination; the Assessor will then have no room to complain!

2.2 Describe the appearance of healthy brood and, in contrast, the appearance of larvae, brood pattern and cell capping that will require further investigation.

2.2.1 The appearance of healthy brood and how it differs from diseased brood or chilled brood.

Brood means all stages of the brood from eggs through larval stages to sealed pupae; simply all open brood and sealed brood. We are concerned here with the determination of healthy, diseased or chilled brood by normal eyesight and not by using optical aids such as magnifying glasses and microscopes. It is not possible to differentiate between a normal egg and one that may be faulty, therefore, the following discussion is concerned with larvae of all ages and sealed brood before emergence.

2.2.1.1 The appearance of healthy brood.
   • Larvae: All larvae, both worker and drone, from the newly hatched egg to the fully grown larvae, just before cell sealing takes place, are pearly white in colour, shiny and lie in the bottom of their cells in a curled up position. It is only possible to see the upper side of the larvae, the lower side is floating in the brood food at the bottom of the cell.
   • Sealed brood: Has convex coffee-coloured cappings made from a mixture of beeswax and pollen ie. porous to allow respiration to proceed normally. Because the cappings contain pollen they are of a dull matt appearance with no trace of shininess.

Diseased and chilled brood will be directly compared with the norm as outlined above. It is extremely important that all beekeepers are capable of recognising healthy brood and thereby being able to quickly recognise disease problems while undertaking normal colony manipulations.

2.2.1.2 The appearance (signs) of diseased brood.

   • Larvae: All diseased larvae eg. stone brood and EFB are coloured and distorted, unlike the shiny, white and circular appearance of healthy larvae positioned in their cells. There are three exceptions. The first is AFB where the larvae before sealing appear to be perfectly normal; the changes happen when the larvae die after the cell is sealed. Similarly in the case of sac brood the sealed cell is uncapped to reveal a larva flattened in shape with an upturned head (known as the Chinese slipper effect). In the case of chalk brood the unsealed cell reveals a mummified larva.
   • Sealed brood: The appearance of the cappings associated with diseased brood is concerned with AFB, chalk brood and sac brood. Any cappings which do not conform to those of healthy brood will have some or all of the following characteristics: sunken or concave cappings, perforated cappings, discoloured cappings and cappings with a darkened damp appearance. Often the cappings are removed entirely revealing the dead larvae or pupae before the bees have had time to remove them. It is instructive to uncap sealed brood around the cell of a dead larva eg. chalk brood, as it often reveals other larvae dead in their cells before the bees have uncapped them. The bees are able to determine whenever there is a dead larva in a sealed cell.

2.2.1.3 The appearance (signs) of chilled brood.

Chilled Brood is brood in all stages which is killed due to exposure to low temperatures. All stages means from the hatched larva to the sealed pupa and for this reason it is very easy to diagnose as no other diseases kill brood of all stages in one fell swoop. It will be clear that it is not a disease but a condition, as no pathogen is involved. Dr.Bailey states that unsealed larvae can survive several days at room temperature of c. 65°F (18°C), so the temperature drop must be quite severe or prolonged to kill them in a colony. We have never seen chilled brood in a colony but have produced it artificially in the refrigerator.

2.2.1.3.1 Causes.

   • When a colony is approaching starvation (there is no carbohydrate to convert into heat

energy).
- Due to spray poisoning (many bees lost).
- Stated in many books to be due to mishandling by the beekeeper (opening a colony for too long in low temperatures) which we don't believe.

2.2.1.3.2 Signs.

- Brood of all stages dead.
- Dead brood at the periphery of the brood nest.
- Some of the capped cells may be perforated.
- Larvae turn grey and then to black in colour and remain shiny though they are discoloured. This is a very important sign and is discrete to chilled brood only. A very positive sign.
- In the later stages a black scale is formed in the cell which is easily removed by the bees.

2.2.2 The appearance of brood patterns that require further investigation.

The final condition associated with brood patterns is the difference between the brood of laying workers and a drone laying queen. In this sub-section we include the treatment of these conditions as it is not addressed elsewhere in the syllabus (it is not much good being able to recognise it and then not knowing what to do).

2.2.2.1 The signs (detection) of laying workers and description of why they occur.

2.2.2.1.1 Detection of laying workers:

a) Drones in worker cells (typical raised domes).
b) Drones produced in this way are small and abnormal (stunted).
c) Laying pattern is scattered and haphazard (cf. drone laying queen which is compact and orderly).
d) Colony endeavours to build charged queen cells (note: this can happen with drone laying queen but is unusual).
e) Workers generally lay more than one egg/cell.

2.2.2.1.2 In the absence of the queen and brood there is an absence of pheromones from the queen and from the brood. These pheromones, particularly that produced by the queen, inhibit development of a worker's ovaries. Workers' ovaries develop in the absence of these pheromones and some workers start laying after about 21 days in the queenless state.

2.2.2.1.3 Cause of colony having laying workers:

a) Queenlessness.
b) Inability to produce emergency queen cells (no fertilised eggs).

2.2.2.1.4 Treatment: it is generally agreed that little can be done except to shake the colony out near a strong stock and let the bees take 'pot luck', always providing that the bees are disease free.

The following points are pertinent:

a) Difficult (impossible?) to requeen; a colony usually kills an introduced queen.
b) Bees are mostly old and of little use to another colony.
c) If they are united to a queenright colony it has been found that there is the likelihood of them killing the queen of the colony to which they are united.
d) Experiments conducted in France in 1989 on the introduction of queens to colonies of laying workers by dipping the queen in royal jelly and water (70% and 30% respectively) are claimed to be a successful treatment. We doubt the validity of this claim.

2.2.2.2 The signs (detection) of a drone laying queen and the causes for this failure.

2.2.2.2.1 The visual signs:

a) Unmistakable worker cells with drone cappings (raised).
b) Presence of a queen (actually seen).
c) Drones produced are small and abnormal (stunted).

2.2.2.2.2 During the season:

a) Queen produces small areas of drone brood in the middle of large patches of worker brood.
b) As the season progresses, worker brood becomes less and drone brood increases.
c) Because some worker brood remains, it is clear that a queen must be laying.
d) Eventually there will be nothing but drone cappings. At this stage the colony will be reasonably large.
e) Drones are smaller and the abdomen stunted.

2.2.2.2.3 In the spring:

a) At the first examination of the colony there may be one or two frames of drone brood only (no worker brood present).
b) Is it a drone laying queen or laying workers?
c) If a queen (drone layer), the laying pattern will be orderly i.e. compact patches of brood with very few empty cells.

2.2.2.2.4 Possible causes for a queen becoming a drone layer:

a) Shortage of sperm - inadequate mating or due to age of queen.
b) Physical inability of queen to fertilise eggs correctly.
c) Genetic fault.

2.2.2.2.5. Treatment:

  a) requeen or
  b) unite after removing old drone laying queen.

2.3 Describe the procedures taken to avoid or reduce the transmission of infectious diseases and demonstrate that these are followed. Spare equipment used by the Candidate will be examined.

2.3.1 Hygiene.

We are again at a loss to understand the requirements in this part of the syllabus noting the content of Sections 1.9 and 2.1 above. These sections adequately cover the requirements for minimising the transmission of infectious diseases.

Perhaps it might be as well to stress that bees carry a host of diseases and swarms and strange bees should not be brought into a fully functional apiary; they should be put in quarantine until such times as they are proved to be fit and healthy.

No colonies or nuclei should be bought or sold without a certificate signed by a suitably qualified and competent beekeeper (a BBKA Master Beekeeper) that the bees have been fully inspected and are free from disease or annotated that disease conditions have been observed (eg. chalk brood, varroosis, etc.).

2.3.2 Spare equipment.

It goes without saying that any spare equipment must be in impeccable order for the examination and the Candidate must have a sound knowledge of what it is used for and how it should be used. Failure should not occur but, as examiners, we have known candidates to fail in this area of the assessment.

2.4 List the reasons for comb renewal and demonstrate the procedures adopted.

2.4.1 The reasons for the systematic renewal of comb are as follows:

  • Minimise disease by removing the associated pathogens on the old comb.
  • Mal-formed and uneven comb to make manipulations easier.
  • Damaged comb (eg. where old queen cells have been removed) to maintain maximum area.
  • Comb with an excessive amount of drone cells.
  • Comb distorted by drones bred in worker cells (eg. drone laying queen or laying workers).
  • Pollen clogged comb.

2.4.2 In order to demonstrate that comb is being renewed regularly it is best to show the Examiner frames in the hive which have been marked with a permanent marking pen with the renewal date.

2.5 Discuss the progress of the colonies as described in the Record Book and the intentions for the rest of the season.

This is a straightforward instruction of what is required for the assessment. See Appendices 2 and 3. We have studied the prospectus in connection with the record book and the hive record cards as shown in the BBKA leaflet 'Maintaining Beekeeping Records'. Both are discussed in the following paragraphs.

2.5.1 The Record Book.

We believe that the important aspects of this are:

   • The requirement for a continuous record for at least one season.
   • The use of a quantitative method of assessment.

It is of little use deciding in the spring to attempt the examination if there are no records of the previous year to be assessed. The decision to put oneself forward for assessment is a long term commitment. A quantitative method of assessment simply means putting some numbers to the item to be recorded in the log; these are illustrated in the BBKA pamphlet.

2.5.2 The BBKA leaflet 'Maintaining Beekeeping Records'.

We have drawn the attention of BBKA to one major error and two minor comments in the leaflet that was published in April 2000 which are as follows:
a) The item 'Room' for stores measured in the equivalent of super frames available should read 'Room' for bees. Supers are for bees and not for honey; if these supers are eventually filled with honey then this can be considered a bonus. It is simple to postulate a situation which illustrates the fallacy of providing room for honey and not for bees and which has happened to us on many occasions. Come the June gap with a large colony filling its empty supers with bees awaiting the main flow; the colony needs feeding to keep it alive and at full strength yet it has more than enough room for any honey it may gather.
b) Under stores we believe that the actual amount in pounds weight or kilograms should be recorded and not an equivalent number. See section 4.6 in our Study Notes for the Basic Examination and section 1.14.3 in our Study Notes for Module 1.
c) Under 'Feed' there is no such thing as light or heavy sugar; these terms should be replaced with thin and thick syrup.
If you have a more recent edition of the leaflet, then ignore the above comments. Enough said, but do remember that when a colony is normally being inspected a large percentage of the foraging force is away from the hive.

2.6 Review the age of existing queens and plans for their replacement. Describe how replacement of queens is carried out.

Opinions will differ and methods will differ on the replacement of queens and how this is done. Much will depend on the queens and the queen rearing programme which is adopted. We describe below the way we go about this aspect of management which has stood us in good stead for many years. Additionally, we look at the hobbyist beekeeper in the United Kingdom in relation to enunciating a procedure for the replacement of existing queens.

2.6.1 The hobbyist beekeeper in the United Kingdom.

There is no doubt that it is essential to requeen the stocks in every apiary from time to time; failure to do so will result in inferior bees that become progressively more aggressive with less and less vitality associated with diminished honey crops. An increased tendency to swarm is also highly likely. The change in character takes place very quickly in about three years. It has happened to us during a period when we could not devote the attention to our bees that is normally required; so we speak from experience. Many beginners start off with a reasonable strain of bee and because they have not learnt the skills of breeding give up because their bees become unmanageable.

Buying in queens is one solution but in our opinion an unsatisfactory one considering the scenario in the UK. A few queens are brought in from New Zealand each year (1,348 in 1995) by commercial or semi-commercial beekeepers but these queens are in a poor state when they arrive after a long flight. Invariably the attendant workers are infected with Nosema making it highly probable that the queen is also infected. We tested the workers from a batch of 12 queens which arrived from New Zealand in April 1994; all had Nosema (they may also bring in unknown viral diseases or parasites eg. *Tropilaelaps clareae*).

Rearing one's own queens is the only remaining solution and this means rearing from mongrels. Breeding from mongrels is like going to the casino; it is a game of luck but there is no alternative. Considering all the characteristics only two or three can be selected and in a small number of colonies the selection of a breeder queen is often self-evident. The characteristics that we use are first and foremost good temper; this is an absolute must irrespective of other desirable qualities. The second is selecting from a non-swarming stock which of course is good tempered and finally if there are two with the foregoing qualities then maximum honey yield will take preference. It may sound curious but honey is not a prime requirement, but regarded as a spin off, as far as we are concerned and we think this will apply to nearly all hobbyist beekeepers. This is why a honey cooperative will not be successful; if the hobbyist has too much honey to sell he will reduce the number of hives that he keeps.

We maintain approximately the same number of nuclei as we have colonies for honey production. These nuclei are maintained throughout the year and relieved of brood if required to stop them 'bursting at the seams' during the season. Queens are reared in May and the new queens mated in the queenless nucs. From the time that they are mated until the following March their progeny is assessed. All our requeening is undertaken in March with the queens raised in the previous May. The old queens are put back into the nucs to keep them going until queen rearing starts again in May. This method of selection generally requires an average of between 10 and 15% of our queens being rejected because of temper, therefore, some of our colonies are not requeened each year and some queens have to survive a further year. Here we must emphasise that about 10% of our colonies supersede each year so it is only a small number of colonies that are not requeened annually. Therefore, our advice is to requeen every year with the proviso that any queen can be allowed to head a colony for a second year.

Our method of selection means that we have to put up with a lot of unfavourable traits, propolis and nervousness probably being the worst. We have bees that can be handled without gloves and these are bred from mongrels that are likely to have bad temper. We have used New Zealand queens and have found that invariably the first cross with the local riff-raff is good tempered and often the second cross also. After that the strain deteriorates badly. For the average beekeeper the purchase of such a queen every two years and using her as a breeder has one overriding advantage and that is a good tempered bee is virtually ensured.

In summary, the number of nuclei = the number of productive hives, queen rear every May, from then on assess the queens for temper, cull the unsatisfactory ones and requeen every March at the beginning of the season.

2.6.2 Description of how the replacement of queens is carried out.

It is not clear what is intended in this section of the syllabus. Is the method of queen rearing to be described together with introduction methods or is it just the actual replacement or queen introduction? We must assume that it is the former as queen introduction is addressed in Section 2.12 later.

2.6.2.1 General considerations. Every beekeeper should rear his own queens, whether he maintains one or two colonies or more. We believe that spare queens should also be available in case of an emergency. With these two thoughts in mind a few criteria can be developed, as follows:

a) If spare queens are to be available throughout the year, then they will have to be reared and kept in overwintered nuclei. The emergency may occur in the spring (eg. a drone laying queen is found in one of the stocks) when it is impossible to raise another queen because of the absence of drones, at that time of the year, for mating.

b) One of the biggest problems in UK beekeeping today seems to be the large number of colonies that are bad tempered. If a queen is overwintered in a nuc it is very easy to determine whether her offspring are suitably good tempered for use in a large colony. If not, the queen can eventually be culled. Bad tempered nucs are easy to handle; bad tempered colonies are very difficult to handle and a nuisance to the beekeeper and his neighbours. Queens can be reared from a good tempered line (as they always should be) but due to the mating a bad tempered strain can result. Mating is out of the control of the beekeeper and he can only monitor the end result. We consider that all queens should first be tested for temper before introduction into the honey producing unit. If all beekeepers followed this advice, their beekeeping would become more enjoyable, neighbours would not be stung and the amount of personal protection could again be reduced to a simple veil.

c) Any queen rearing should be planned and this must include consideration of the following:

1. Timing - ready to start at 2nd or 3rd week in May in the UK.
2. Selection of the breeder queen.
3. What method of larval transfer is to be used?
4. Selection and preparation of the cell building colony.
5. How many queens are required?
6. Mating nucs.

Each of the above points will be examined in some detail with a view to evolving a suitable approach for the small hobbyist beekeeper. However, before setting out to rear any queens it is absolutely essential to fully understand the life cycle and natural history involved from the laying of the egg by the breeder queen to the new queen actually laying. Once the timing schedule has started, there can be no variation; come hail or shine, each operation has to be done on time. Opening a colony under an umbrella is not too easy on one's own and in this sort of situation good tempered bees are definitely preferred.

2.6.2.2 How many queens?

There is no indication in the Prospectus, of how many queens are to be reared, and the wording in the syllabus is very ambiguous where in Sections 1.2 and 1.3 it states that 3 colonies and 1 mating nuc are to be made available for the assessment. It would be better to define the number of queens required. 5 to 10 queens would, in our opinion, be a small number and suitable for an apiary of the same number of hives. 10 to 50 queens would be a medium number and 50 to 500 or greater would be in the commercial queen rearing class. The number is important because to rear good queens the larvae have to be well fed and the more there are the bigger and stronger the cell building colony has to be. There is, of course, a finite limit to the capabilities of the strongest colony in respect of the number of queen cells (QCs) it can successfully raise. Depending on the method used, so the skill and experience of the beekeeper will influence the success rate. This is particularly true if grafting is adopted. It is therefore prudent to allow a safety factor for this 'success rate' and 50% success is quite realistic for the moderately dextrous operator. So if you want 10 aim at 20. It is always best to use simple methods until more demanding techniques are learnt; simple methods that work will be quite acceptable to the Examiner. Additionally, queens will be lost on mating flights and some queens may not mate properly and become drone layers. It is extremely unlikely that a 100% success rate will be achieved no matter how skilled the beekeeper may be.

2.6.2.3 Timing.

The critical factors in queen rearing are:

a) Mature drones are required for mating (a minimum of 12 days after emergence).
b) Optimum conditions are those associated with the time that colonies swarm naturally.
c) A flow is virtually essential but this can be simulated by feeding.
d) There must be an abundance of nurse bees for feeding the larvae.
e) Royal jelly is synthesised from pollen and this is required in quantity.

Considering all the above factors, they indicate that, in the UK, the month of May is generally the time that they all occur naturally. We always aim to have the cell building colony ready by the 2nd or 3rd week in May. If bad weather occurs the start date can slip a few days but once started there can be no slippage. At the first inspection in the spring, the cell building colony should be selected (the breeder is likely to have been chosen the previous season) and this colony must be built up until it is teeming with bees and brood by mid-May. Queen mating nuclei will be required 9/10 days after larvae are introduced into the cell builder and provision for making these must also be included in the timing schedule.

2.6.2.4 Selection of the breeder queen.

This is the queen which is selected for her good characteristics which hopefully will be reproduced in her progeny, daughter queens. There are many characteristics, however, and because of the importance of this matter we will reiterate those that are important for the hobbyist beekeeper which, in our opinion, are as follows:

• **Temper** must come at the top of the list. It is essential for the breeder queen to come from a colony which produces good tempered bees which can be handled without gloves and with a minimum of smoke. If this requirement cannot be complied with, it will be necessary to buy in a queen of known good temper to breed from. Having ensured the temperament of the breeder queen it does not follow that success is also ensured; the mating of the young virgins finally casts the die.

• **Nervousness.** This trait is exhibited by bees that move quickly and run all over the comb during inspection, finally clustering in a bunch at the bottom of the frame and falling off leaving the comb virtually bare of bees. It is extremely difficult to find a queen in such a colony and often we have found that they are usually difficult to requeen by normal introduction with a Butler cage. When a frame is removed from a colony during a normal inspection, the bees should remain quiet and completely cover the frame while it is out of the hive. If they can't meet this test, then don't use the queen as a breeder.

• **Swarming.** Do not breed from a strain which swarms prolifically (producing a large number of queen cells). The Heath bees brought into this country from Holland after the I.O.W. disease (Acarine now Acariosis according to Brussels) are inveterate swarmers and the strain is still with us in our mongrel bees. Some years ago we acquired a queen from a fellow beekeeper which passed on the above traits but turned out to be a swarmer when we bred from her; it was impossible to stop them swarming and we had to dispense with them in the end.

• **Disease.** Some strains of bee are more resistant to a particular disease than others. It goes without saying that the breeder queen should come from a colony with a disease-free record.

• **Fecundity.** A prolific egg layer means a large colony and lots of bees which in turn means a larger foraging force for honey production. It is to be noted that the very yellow coloured bees (Italians originally) now brought in from New Zealand are very prolific and convert every bit of food into bees. They often require feeding when other strains can survive on their own resources and they tend to be prone to Acarine infection; which is a pity because they are delightful bees to handle.

After a season has been completed, most beekeepers are well aware of the colony which is headed by a queen with the least undesirable characteristics (unlikely to be the most desirable!) and selection is therefore very easy. One characteristic which is not mentioned above is the tendency to collect propolis in large quantities. With one or two hives it never seems to matter very much but when 20 or more colonies are 'propolisers' it does become a little irksome during manipulations.

2.6.2.5 Selection of the cell builder.

The cell building colony will receive the 12 to 24 hour old larvae and build them into queen cells after a period of queenlessness. The colony has to be very strong and teeming with bees and is to be opened at fixed times irrespective of the weather. It is therefore important that this colony should also be of good temper. A further factor for consideration is whether to work a single brood chamber or a double one. With hives using BS frames (eg. British National) a double brood chamber system is likely to be the most popular format for most beekeepers. For the hobbyist beekeeper it is probably more convenient to do his queen rearing at home rather than at his out apiary. In such a case, there could be advantages in initially preparing the cell builder at the out apiary and moving it to the home apiary on to a site just previously occupied by another colony so that the cell builder is further reinforced by the flying bees from the stock moved in the home apiary. It should be noted that the cell building colony and the breeder can be one and the same stock if required.

2.6.2.6 Larval transfer.

There are a variety of methods of introducing larvae from the breeder queen's colony to the cell builder; some of the more well known approaches are listed below:

a) Grafting (something of a misnomer) is physically transferring the larva with a small tool (a grafting tool, another misnomer) from a worker cell in the breeder colony to an artificially made queen cup which is then put into the cell builder. Specially made tools are sold made of metal with a minute spoon at one end. They are expensive and all the more experienced queen rearers that we know use something to their own liking such as a chewed matchstick, a small camel hair brush moistened with saliva between grafts, etc.
b) Double grafting - a larva is transplanted into a queen cell and then one or two days later before it is sealed it is replaced with a new young larva.
c) Punching out the whole worker cell containing the egg or young larva. The punched out cell is then attached to a suitable bar in a standard frame for insertion into the cell builder. Examples are the Barbeau and the Stanley methods. The advantage is that the actual larva is not touched and therefore reduces any likely damage to the larva and hence failure.
d) Transfer of a whole frame of eggs and larvae from breeder to cell builder direct. There are variations on this eg. Miller method of cutting the comb in 'Vs', the line of the cut being adjacent to larvae of the right age. A frame can be put over the top of the cell builder horizontally (about 1 inch above the cell builder frames) allowing the cells to be formed vertically from the lower comb face (popularly called the Australian method supposedly because the frame is the wrong way round!).
e) A more recent approach is to use the Jenter method. These methods prevent physical damage to the larvae and require little dexterity as the larvae are selected by the bees.

A simple method of raising 5 to 10 queens would certainly rule out grafting methods which are essential for raising large numbers of queens and require good eyesight, a steady hand and some experience to have a fair degree of success. When we undertake grafting, we do 10 grafts each to provide a total of 20 in 2 separate rows to check our own dexterity one against the other (so far there has been no clear winner or loser!) which is a useful monitor.

2.6.2.7 Mating nucs.

Provision will have to be made for the mating nucs and be part if the overall plan. Because of the vagaries of the UK weather it is probably better to have the nucs made up prior to the day that the ripe queen cells are transferred. If it is pouring with rain the transfer is quite enough to do.

It is appropriate to discuss mating nuclei in more detail in this part of the syllabus as the size and the environment inside the nuclei are very important if first class queens are to be the end result. Generally there are 3 types of nucleus for queen mating; starting with the largest using say 3 to 5 BS frames, a mini nucleus using small frames such as the commercially available 'Apidea' constructed in high density polystyrene and finally the smallest, a micro nuc, with one tiny frame perhaps with cells on one side only as described in 'Micronucs' by John Atkinson*. Clearly the different size nuclei will accommodate different populations of worker bees.

Scientific work has been undertaken on the migration of the spermatozoa into the queen's spermatheca after mating. Under the 'right' conditions this takes approximately 40 hours after which the queen starts to lay. It has been established also that the time elapsed between emergence and mating is dependent on the weather outside the hive and the environment inside the hive.

In 1982 Woyke and Jasinski undertook further investigations on 4 sizes of mating nuclei as follows:

  a) Micro nucs, 1 comb, 150 workers.
  b) Mini nucs, 2 combs, 350 workers.
  c) Mini nucs, 4 combs, 750 workers.
  d) Standard 5 comb nuclei, number of workers not stated (3,000 to 4,000, our estimate).

Their conclusion was that a minimum of 350 workers is necessary in order for a normal number of sperm to migrate into the spermatheca. In 1990 after further work they reported that the time from insemination of the queen by the drone to the start of oviposition (egg laying) is determined by the number of workers in the mating nucleus:

  The greater the number of workers, the shorter the time to oviposition. In a standard 5 frame nucleus it takes 5 to 9 days compared with the latest start in a 150 worker nuc where it took 24 days.

Further work was undertaken quite recently by L.A.M.Hassan at Cardiff for a postgraduate degree. He used 3 types of nuclei namely, micro nuc, mini nuc and normal (presumably BS frames 3 or 4). Observation of 92 virgins was undertaken with the following results:

* See chapter 5 of Atkinson's "Background to Bee Breeding" NBB, 1999.

Pre-mating periods -	shortest - normal nuc
	longest - micro nuc
Pre-oviposition periods -	shortest - normal nuc
	longest - micro nuc
Queens lost -	more in micro nucs than any other type
Deaths of queens -	greatest in micro nucs
Abscondings -	greatest in micro nucs, none from normal nucs
Spermathecal volumes -	greatest in normal nucs (ie. greatest number of spermatozoa in spermatheca)

We believe the above findings to be extremely important when considering the type of nuclei to use for introducing ripe queen cells and the subsequent mating of virgin queens. We have never used micro nucs but gave up using mini nucs many years ago because of unsatisfactory results. Our reason for abandoning them was stress which caused Nosema and which was always present in mini nucs despite using Fumidil 'B' whenever they were fed. Nowadays, we use only 4 and 5 frame nuclei using British National or Commercial frames.

The mini nucs were developed and designed to provide an economy of bees and may be satisfactory in climatic conditions which are more favourable than those normally experienced in the United Kingdom.

2.6.2.8 A simple method of raising 5 to 10 queens.

Before attempting any queen rearing whether it be on a small or large scale by commercial beekeeper or hobbyist, a timetable must be drawn up and, once started, adhered to without any deviation. The timetable is essentially a simple PERT diagram (progress evaluation review technique) well known as a management tool in any project management. It will define precisely what has to be done and at what times; there must be no deviation from the critical path.

This simple method is based on double brood chambers with BS frames and using the breeder colony as the cell builder; 5 to 20 queen cells may be expected, the exact number will depend very much on the genetic make up of the strain of bee. The steps in the process are listed below:

• Prepare the cell building colony for queen rearing, aiming to have the largest population and brood by mid May when there are likely to be mature drones for mating. The stock would be 2 full brood chambers, queen-excluder (QEx) and two to three supers.
• Draw out a programme with actual calendar dates incorporated below the natural cycle from egg to emergence; ie. 3 days for the egg to hatch, 5 days as a larva before the cell is sealed with the emergence occurring 15/16 days after the egg was laid.
• Re-arrange the colony for queen rearing using a spare empty brood box. Into the spare brood box put all the frames of open brood and eggs and as much sealed brood and pollen as possible but with no queen. The queen and the balance of frames are placed on the floorboard with QEx over. Add the supers and the box of brood with no queen with crown board and spare eke and feeder (1 pint size is adequate). Feed approximately 1 pint per day. As a result of this rearrangement all the nurse bees (or most of them) will join the brood in the top box and in 24 hours queen cells (QCs) are

likely to be started; this must be checked. If a cell is not started, insert a screen board with entrance under the top brood box. There can now be no possibility of any queen substance reaching the bees in the top box and QCs will be started.

• After 3 days examine the top box and destroy all sealed QCs (these will have been built on larvae 2 or more days old). Count and leave all other open QCs which should be sealed in another 1 or 2 days time. There are two problems with this system of queen rearing which are as follows:

> 1. When destroying the sealed QCs most of the bees will have to be brushed off the frames to ensure that all the sealed ones are destroyed. Do not shake your future queens.
>
> 2. One can never be quite sure when the remaining open cells are sealed. It is therefore very important to mark up the programme and be ready to remove the ripe QCs on day 14 approximately 2 days before the theoretical emergence. If one or more emerge earlier than your calculated day 16, it will not ruin the whole programme; the ripe queen cells will be safely in their mating nucs and the first one out will not be capable of destroying all those in the cell builder.

• Four or five days before the ripe QCs are to be distributed, nucs can be prepared and left queenless for four or five days if they are permanent nucs. This means de-queening them and then destroying all emergency QCs just before introducing a ripe QC on day 14. All frames must be shaken to ensure no scrub QC is left in the nuc.

• On day 14 new nucs can be made up with only sealed brood (no eggs or open brood) and the ripe QCs distributed to them.

• The top box can be left with one ripe QC and when this new queen is mated and laying the top and bottom boxes can be united after removing the old queen in the bottom box. Alternatively, the top box can be used to make 3 good nucs for some of the ripe QCs.

2.6.2.9 A variation on the above which is more precise.

• Prepare the cell builder as stated earlier. The colony must be thriving, opulent and disease free. It can be on a single or double brood chamber; a single one is easier to handle.

• Dequeen the cell builder. The queen can be temporarily kept in a nuc. 7 days later destroy all the emergency queen cells by shaking every frame to ensure that it is hopelessly queenless.

• The following day introduce grafted larvae of a known age (two rows of 10 on two cell bars on one standard size frame). Ensure it is in the middle of the brood chamber with frames containing ample pollen on either side of the cell bars. Because the age of the larvae is known there can be no mistakes in the timing.

• Distribute ripe cells to hopelessly queenless nucs on day 14.

2.6.2.10 Methods of producing larvae for the cell builder.

The methods are legion and it would be inappropriate to discuss them all in detail here. For further reading the following list will suffice: Miller method, Alley method, cell punching methods eg. Stanley, Barbeau, etc., Jenter method, Australian method (horizontal frame with eggs over queenless stock), etc.

2.6.2.11 Methods suitable for larger scale queen rearing operations.

To complete the discussion let us consider briefly large scale queen rearing. The limitations on the number of queens reared or the number of cells prepared by a cell building stock is finite and depends on the size of the colony. As a guide we consider that with our mongrel bees a single brood chamber (11 BS frames) is capable of raising 5 to 15 queen cells containing well nourished larvae. If a greater number is required then the size of the colony can be assessed on a pro rata basis or more than one colony can be used. A double brood chamber with 22 BS frames can produce 15 to 30 cells.

At Buckfast Abbey, Bro.Adam used Buckfast Dadant hives with double brood chambers and 20 frames. Such a colony was expected to nurture 60 grafted larvae. It must be emphasised that for the best queens the larvae have to be well nursed (fed) and this requires the colonies to be teeming with bees.

Nucs are required for receiving ripe queen cells in the same way as described for rearing small numbers of queens and it will be clear that, if heed is taken of the requirement for a standard nucleus, it is likely that preference will be given to maintaining permanent nuclei in a major queen rearing establishment. Making up hundreds of nuclei near the time the ripe queen cells require to be distributed becomes a mammoth task unless the nucs are very small. In New Zealand, Australia and USA where the largest queen rearing operations are undertaken, mini nucs are the order of the day. Due note should be taken of the more favourable weather patterns in these places but also note our experience with imported queens from New Zealand with Nosema.

When queens are sold provision has to be made for the bees in the queenless nucs by combining two or three of them with a common queen to maintain the population. Otherwise resort must be made to keeping queens in bank colonies.

2.6.2.12 Queen banks.

Colonies in which queens are stored are called queen banks and are essentially the same for both virgin queens and for mated queens. It is possible to store from about 10 to 100 queens in these bank colonies. The bank colony must be queenless and provided with a continuous supply of new bees every week (say of the order of 1 to 2 pounds or 5,000 to 10,000 bees) by introducing frames of emerging brood. The queens are kept in individual cages in special frames interspersed between frames of emerging brood and frames of pollen. The colony is continually fed with syrup dosed with Fumidil B to prevent the spread of Nosema (the colony is under stress and Nosema is a stress disease).

Virgins can only be kept for about 3 or 4 days and then must be introduced to mating nuclei to allow mating to take place. Mated queens can be kept much longer and it is normal for the large commercial queen rearing enterprises to employ queen banks. It is important to note that very young bees are quite tolerant to foreign queens but become increasingly less tolerant as they become older. Intolerant workers will damage caged queens through the wire mesh by biting the queen's antennae, legs, etc.

The practice is, in our opinion, an unnatural one and is best avoided if possible.

2.6.2.13 Other points of interest.

• Queens are expensive to buy and after travelling they are not in the best condition for introduction. They are only available at the wrong time of the year if produced in the UK for use in UK. The beekeeper with a small number of colonies can produce queens the equal of those at top prices with the material he has available on his own doorstep if he is so inclined.
• In order to get rid of bad temper it is essential to maintain overwintered nucs to test the new queens from the time they are reared to the following March when they should be introduced. Our own experience with mongrel strains amounts to culling approximately 10% of those reared.
• For 20 stocks we maintain approximately 15 permanent nucs for re-queening purposes. The nucs are virtually self supporting.

2.7 Describe the methods taken throughout the year to monitor and control Varroa to non-damaging levels. Demonstrate the use of Varroa control equipment in the apiary. Examine a brood chamber and floor for Varroa. Demonstrate the use of comb for trapping mites in drone cells.

2.7.1 An outline account of the signs of Varroosis.

Varroosis is the disease caused by *Varroa destructor* formerly *Varroa jacobsoni*, (a mite of the class Arachnida).

Varroosis not only affects the sealed brood where it breeds but it also lives on the adult bee and feeds on its haemolymph at the intersegmental membranes usually on the ventral side. Therefore, it can be classed as both a brood disease and an adult bee disease.

Signs:

It is unlikely that any sign of Varroosis will be apparent until the colony has been infected for about 3 years. The first indications are likely to be a general weakening of the colony.
It is very unlikely that the mites will be seen on the adult bees as they generally inhabit the intersegmental membrane area on the ventral side of the abdomen. The most positive sign is by knock down test using Bayvarol or Apistan and collecting the knocked down mites on a paper insert below a Varroa screen.
As the *V. destructor* breeds in the sealed brood cell (drone preferred) these also cannot be seen except by opening the cells and conducting a systematic search.
At an advanced stage of infestation (say 2 years +), when the infestation has become heavy, underweight workers will be produced due to malnutrition and then deformed worker bees are likely to appear in the colony. Such bees may have stunted bodies and/or deformed bodies, wings and legs. There are some good colour photographs in the MAFF 1992 pamphlet (PB 0925 on the back page) of these deformities. There are likely to be 3 or more mites per cell at this stage and there will also be signs of neglected brood and the spread of secondary infections.

It has been shown in Germany but not in UK that *Varroa destructor* is a vector for Acute Bee Paralysis Virus and it is likely that signs of ABPV may be seen before other signs of *V. destructor* infestation.

When the colony is at an advanced stage of infestation there is every likelihood that it will collapse suddenly and die out most probably at the end of the summer. The mechanism may be that as the queen reduces her laying after the main flow there are less brood cells for the female mite to enter and more than one mite per cell results. With multiple females occupying one worker cell deformed bees are inevitable, leading to rapid colony collapse. This emphasises the importance of detecting the mite at an early stage of infestation in order that the colony may be saved in time.

If the beekeeper detects an infestation he is still obliged by law to report the situation to DEFRA (formerly MAFF) despite the futility of the exercise.

It is incumbent on every beekeeper to make a regular search every year for this scourge and treat it accordingly.

2.7.2 Methods of detecting Varroosis.

There are now a wide number of methods for detecting *Varroa destructor* infested colonies; these are:

Examination of hive debris on the floorboard in early spring. This method has been advocated by the NBU for some years and at the time of writing they will still undertake the examination free of charge. The method has its limitations and is not wholly reliable particularly when the infestation in the colony is light. The debris from all the floorboards in an apiary is collected and bundled together and sent to the NBU* with the name and address of the beekeeper and his apiary. After examination the NBU send back a report stating the results. If the sample comes back with a negative result and a light infestation has been missed it will be of little consequence on colony performance during the coming season. The method must be quite attractive to many beekeepers until such times as it is discontinued or until the apiary concerned has been diagnosed positive.

Uncapping brood, particularly drone brood (pupae at the pink eye stage) with an uncapping fork during regular colony inspections. Care must be taken when examining the larvae for reddish coloured mites; in the early stages of development the nymphs are virtually colourless and translucent. Special frames of drone foundation can be inserted in the brood chamber for this purpose. Note that this is also a manipulative method of Varroa control without recourse to chemicals.

Bayvarol test using a Varroa screen with paper insert below together with one strip of Bayvarol (flumethrin) or Apistan (fluvalinate) both of which are synthetic pyrethroids. Bayvarol and Apistan are the only medicament approved for use in the UK at the time of writing and should be left in the colony for 24 to 48 hours. The strip and insert are then removed and the insert examined with a magnifying glass for dead mites. The medicated strip is inserted between the centre frames of the

*NBU, Sand Hutton, YORK, YO4 1LZ

51

cluster and can be introduced without moving the frames and disturbing the colony. It is the most effective detection method available at the present time. The strip can be re-used to test other colonies in the same apiary noting that other diseases can be transferred from one colony to another on the same strip. When these strips are used for treatment, they are left in the colony for 5 to 6 weeks, ie. the strips have a working life, according to the manufacturers, of 6 weeks used in an average sized colony. For testing purposes it would be prudent to use the single testing strip in the colonies for say no longer than 21 to 30 days of actual use before it is scrapped thus ensuring a reasonable surface density of medicament to knock down any mites that may be present.

Tobacco smoke test. We will not describe this test as much has been written about it that it may be assumed that the method is universally known. Compared with the use of a Bayvarol or Apistan strip it must be regarded as a 'stone age' method of testing. We have tried it many times and would not now recommend it. It comatoses a very thick layer of bees on top of the Varroa screen at the bottom of the hive which must put the colony under stress. We have tried varying the tobacco quantities and the types of tobacco but the result is the same. There are reports from Germany that the beekeepers there are abandoning the method because of loss of queens; it is not clear how the queens are lost but the most likely mechanism would be that the queen becomes comatosed and then she is balled before she recovers. We now know that tobacco smoke can stun some of the Varroa mites as opposed to killing them. They subsequently recover and walk off the paper insert and back into the cluster unless the insert is greased with Vaseline or some other form of grease.

Testing a couple of frames taken from a colony during a routine colony inspection, putting them into a nucleus box and smoke testing it. This is a method suggested by DARG (Devon Apicultural Research Group) but, in our opinion, is not a very practical approach to the problem of detection. For the beekeeper with only one or two stocks, each complete colony could be quickly tested and for the beekeeper with 2 or 3 apiaries time will not normally permit such tinkering during routine inspections.

The use of specially designed floorboards and combined Varroa screens and continually counting the number of dead mites appearing on the insert. This really is a monitoring method to know when treatment should be undertaken, whether treatment has been successful or whether re-infestation has occurred rather than a detection method. It is a management method and widely used in Germany. The combined floorboards/Varroa screens are quite expensive and are not likely to be popular with beekeepers unless they make them themselves. The method relies on the natural mortality of the mites. Using the method the German have been using the following data:

> 1 dead mite is equivalent to 120 mites in the colony in summer and 500 mites in autumn or spring. The average mites/day figure is calculated for the whole apiary from say a minimum of 5 hives. The critical infestation point is taken as 10 mites/day giving between 1200 and 5000 mites in the colony. Treatment must be undertaken at this point otherwise tens of thousands of mites may be expected the following year with colony collapse soon to follow.

Detection with Folbex VA. These fumigation strips containing bromopropylate were specially designed for the treatment of Varroa and Acarine. They are not generally available now in the UK (said to be carcinogenic) but can still be regarded as an effective detection method. Available from France.

Detection of Varroa in swarms is an important aspect of practical beekeeping. Since any mites will be on the bees (no brood), then if they are detected the swarm can be treated accordingly. A simple straightforward method is to shake about half a cupful of bees into a jam jar of petrol. The contents are then coarse filtered to take out the bees and then fine filtered through filter paper to reveal any mites if the swarm is infested.

If a colony is infested and it is to be treated with Bayvarol or Apistan, the treatment will take 6 weeks and this should be done when the colony has no honey supers above its brood chamber. Working back 6 weeks from April means that spring testing for Varroosis should start in about mid February in the south of England. We do not think that this point concerning the early start has been fully appreciated by many beekeepers.

All detection is dependent on being able to recognise a *Varroa destructor* mite and not to confuse it with a *Braula coeca*. The physical differences are as follows:

> *Braula coeca*: ellipse shaped c.1 - 2mm with 6 legs, coloured reddish brown. It is a wingless fly. Initially it is white and takes about 12 hours to turn colour after hatching. The head and posterior end of its abdomen are on the ends of the major axis of the ellipse, the legs are on the sides associated with the ends of the minor axis of the ellipse looking down on the dorsal side. Easily seen by eye riding on worker bees and very often the queen is infested. Causes no harm to queen or bees but the larvae spoil capped honey comb with fine tunnels in the cappings.

> *Varroa destructor*: also ellipse shaped c. 1.1 - 1.7mm. with 8 legs, coloured reddish brown the same as the Braula coeca. The legs are on the ventral side and cannot be seen when it is viewed looking down on the dorsal side. It travels 'blunt end first', its legs being on the sides associated with the ends of the major axis. It is an arachnida and is in the spider class in the animal kingdom not the insect class. These mites are difficult to detect as they feed on the haemolymph by piercing the membrane between the abdominal segments on the adult bee and breed in the capped brood cells.

The question that has not been addressed fully in the MAFF (now DEFRA) pamphlets is when should detection of Varroosis be undertaken? Until Varroosis (previously known as Varroasis) is detected in an apiary, we believe that detection should be a minimum of twice a year. Once very early at the end of winter / beginning of spring and later at the end of the main honey flow when supers have been removed. If the specially designed floorboards/Varroa screens are being used the search can be continuous throughout the season looking for dead mites that have died from natural causes. Our own experience with 3 apiaries within a few miles of each other revealed negative results when sample tested with Bayvarol in the spring but all were positive when tested the same way in autumn of the same year.

For beekeepers with large numbers of colonies and more than one apiary, sample testing (say one hive in three in an apiary) would be satisfactory.

It should be pointed out that, at this stage, there is no preferred method of detection. Our own opinion in this matter is that testing with a Bayvarol strip for 24 hours is the most reliable at present.

2.7.3 Monitoring the presence of the mite in colonies.

Once the mite has been detected in the colonies of an apiary, any monitoring process must involve counting dead mites that have died a natural death or counting the mites killed by knock down tests at regular intervals. If mites have been detected in one colony in an apiary, then it is safe to assume that all colonies in that apiary are infested or will be very shortly even if tests on these other colonies prove negative in the initial stages.

To monitor dead mites it is necessary to have a special floorboard with an integral Varroa screen (which are unpopular because of their high cost) or resort to a DIY job with the existing floorboard turned through 180º with a screen above.

The frequency of each monitoring session must be decided and the more often the better for the best results. Whether this is once a week or once a month must rest with the beekeeper and the amount of time that he can devote to the monitoring as part of his management system. Once the number of colonies rise, say above five, the work of monitoring will quickly become very time consuming and tedious.

It is not much use counting mites as a continual monitoring process unless action is taken if the levels rise above a predetermined threshold. For management purposes reference should be made to MAFF leaflet No. PB 3611 1998 entitled "*Varroa jacobsoni*: monitoring and forecasting mite populations within honey bee colonies in Britain". The details in this leaflet have been embodied in a Varroa Calculator which determines when treatment is necessary dependent on the natural mite fall which is supplied free to all beekeepers by the DEFRA. It has not proved to be a popular management tool with the general run of the mill beekeepers but any candidate for this examination should be familiar with its workings. It should also be noted that if the colony has had supers added then it would be unwise to treat until they are removed which is likely to be at the end of the season.

The question must therefore be asked whether there is any purpose in monitoring on a continuous basis throughout the season? We are of the opinion that any monitoring should be regarded as a management operation undertaken when the colony has no supers above its brood chamber unless there is something obviously wrong which is not attributable to any other known cause. The equipment required to provide continuous monitoring demands screens that do not corrode and made of a material that cannot be damaged by the bees themselves.

There has been, as far as we are aware, no recommendations for preferred monitoring methods nor any objectives defined for undertaking it on a long term basis from either DEFRA or the BBKA. We would have thought that some guidance from both on this point would have been desirable.

It is highly unlikely that bee farmers with commercial enterprises will have sufficient time or effort available to undertake monitoring programmes; the economics of running such businesses would preclude it. Most of the hobbyist beekeepers are unlikely to be interested also. Thus any monitoring is likely to be left to the informed hobbyist beekeeper who has an interest in this new disease. Thus the majority of beekeepers in the UK will require control methods to be applied annually or bi-annually that are simple to use and relatively cheap.

2.7.4 Demonstrate the use of Varroa control equipment in the apiary.

It goes without saying that the Candidate must be 'au fait' with the equipment associated with Varoosis. This aspect of the syllabus is very important and should receive the full attention of the Candidate.

2.7.5 Examine a brood chamber and floor for Varroa.

We presume that this must mean being able to identify the mites and not to confuse them with *Braula coeca*. It is generally possible in most colonies these days (as all are infested with Varroa mites to a greater or lesser degree) to identify mites in the colony on the bees. The Examiner will probably ask the Candidate to find one during the inspection. Examination of the natural mite drop on to the Varroa floor during the apiary assessment will need a suitable grid in order to do the counting.

2.7.6 Demonstrate the use of comb for trapping mites in drone cells.

This must be a very easy part of the syllabus by providing about half a brood frame of drone comb in the brood chamber which can be removed and replaced when it is capped, thereby trapping the mites inside the cells together with the drone larvae. The Examiner is likely to ask what you do to remove the larvae and the mites. Hose them out, after de-capping the cells, away from the apiary and let the blue tits do the rest is probable the easiest method of disposal.

2.8 Describe the routine measures taken to look for disease in the colony.

There are 5 routine measures to be taken when looking for diseases which are as follows:

1. Early spring.
2. At every colony inspection.
3. In the autumn before closing the colony down for winter.
4. Annual special inspections for the foul broods.
5. Monitoring for Varroosis.

Each of the above is dealt with in the following paragraphs.

2.8.1 Early spring.

At the first colony inspection (not very early, when floor boards are changed) a sample of about 30 bees should be taken from the colony for microscopic examination for the adult bee diseases. It is important to know where to take the sample to ensure that old bees are collected, that is, from the end frames of the brood nest or trapping foragers exiting or returning to the hive.

Depending on the results the colony should be treated accordingly. Nosema is the biggest enemy. Early in the year one cannot distinguished between spring dwindling (often used phraseology for Nosema) and a scrub queen without recourse to a microscopic examination.

2.8.2 At every colony inspection.

Colonies should be inspected on a regular basis (7 or 10 day inspections) after the first inspection of the year. Firstly to ascertain colony build-up, secondly to ensure that no swarming takes place and finally when preparing the colony for winter. During these inspections there are five items which should always be checked and recorded on the hive record card as follows:

1. Are there sufficient stores to last to the next inspection if there is no income available?
2. Is the queen present and is she laying normally?
3. Is there any sign of disease?
4. Is there sufficient comb space for the queen to lay and for the bees (remember many foragers will be out when the colony is inspected)?
5. Has the colony built up since the last inspection and/or are there preparations for swarming?

At every inspection the beekeeper should be looking for disease in the brood chamber. All disease free and healthy open brood is very shiny and pearly white. The trick is to look for brood that is off colour and dull rather than looking for healthy brood. The only exception is neglected drone brood and addled brood which is defined as abnormal brood but the abnormality is not known; log it and look in the same place at the next inspection and, usually, it has disappeared. If not, then get hold of a Master Beekeeper for assistance.

2.8.3 In the autumn before closing the colony down for winter.

This is virtually identical to the spring inspection and a sample of bees is required to be sent for microscopic examination so they can be treated before the colony settles down for winter and forms a cluster. It is the clustering process which spreads the diseases.

It has now become the norm to treat for Varroosis at the end of the season with Bayvarol or Apistan strips. This will be complementary to the monitoring programme for Varroosis discussed earlier in this chapter.

2.8.4 Annual special inspections for the foul broods.

The two statutory brood diseases AFB and EFB require special attention each year. In a well managed apiary all colonies should be examined once per year specifically for these two diseases and it involves a drastic upheaval in the colony activities for a short while.

ADAS leaflet No. P306 Revised 1982 'Foul brood of bees: recognition and control' should be obtained (free of charge) which contains some excellent coloured photographs of both brood diseases.

• Both diseases are diseases of the brood and there are no signs associated with the adult bees in an infected colony. In order to diagnose either in the field, it is necessary to open up the colony and examine the combs containing brood. To do this properly it is necessary to shake the bees off the comb before examining it, leaving no more than a few bees on the comb. The reason for this is that in the early stages only an odd cell or two will be exhibiting the tell-tale signs. This important aspect of searching for the diseases is frequently overlooked and inadequately expressed in much of the literature. There is a right and a wrong way of shaking bees off combs, the objective is to rid the comb of bees and keep them in the hive (not flying around the apiary); therefore raise the comb slightly and shake it sharply within the brood chamber without jarring the rest of the colony.

• In order to diagnose the diseases in the field it is easier to remember the signs if one has an understanding of the progress of the diseases:

AFB (American Foul Brood): The larva is fed the AFB spores with the larval food. The spores germinate in the ventriculus and the larva dies after the cell is sealed. The germinated spores break through the wall of the ventriculus into the haemolymph and the larva dies of septicaemia * ; then the whole larval form disintegrates, melts down, becomes thick and sticky and finally dries to a hard scale on the lower angle of the cell. During this deathly saga the colour changes from white to black. It is most important to note that prior to the sealing of the cell, the larvae appear to be perfectly healthy.

* Septicaemia - is the circulation and multiplication of micro-organisms in the blood.

EFB (European Foul Brood): The larva is again fed the pathogen, this time a bacteria which is not spore forming as was AFB, which multiplies in the ventriculus by using the larval food and the larva dies before the cell is sealed due to starvation. It dies at about day 3 or 4 before the cell is sealed, so it is quite large when it dies. A dead larva is not sealed by the bees and is removed. During the starvation period the larva contorts into unnatural shapes in its cell and changes colour from a pearly white to cream to yellow to light browny green (colours are difficult to describe in words; any deviation from the pearly shiny white must be regarded with suspicion). When the bees remove the dead larvae, they are removed in one piece. The infected larvae are either there to see or else the signs have been removed.

It is important to look for EFB at the right time because the bees remove the dead larvae. The right time is during the colony build up in the spring when the amount of brood temporarily outnumbers the adult bees nursing this brood. Those familiar with the annual population cycle (see Appendix 8) will know that this occurs about March or April and we would suggest that mid April is a good compromise to search for both diseases.

It is important to stress that, because special attention is devoted to these two diseases, they may not be ignored for the rest of the year. The beekeeper must continually look for them at each and every inspection of the brood chamber.

2.8.5 Monitoring for Varroosis.

This has been addressed in Section 2.7 above.

2.9 Demonstrate the inspection of a brood comb for brood diseases.

Having shaken all (or most of) the bees off the frame into the hive as described in section 2.8.4 above the next step is how to hold the frame to examine it. It should be held at an angle of about 45° degrees towards the light to get the best view of the lower angle of the brood cells to determine whether there are any remnants of melted down larvae in the form of black scale.

It is really very simple when one understands the course of the disease and what happens after the larva dies. No candidate for the examination should fail on this part of the syllabus. Conversely, not being able to demonstrate the method and the reasons behind it would, in our opinion, be the subject of a fail.

2.10 Demonstrate taking a sample for the diagnosis of adult bee diseases.

It is important to understand that we are examining bees that may have adult bee diseases and it is the old bees that are more likely to be infected rather than the young ones. A bee that has just emerged from its cell will NOT be infected with any of the adult bee diseases except maybe Varroosis which is specifically excluded from our present considerations. A recent BBKA Advisory Leaflet No102 'Taking a sample of adult bees', the only one that we know of, unfortunately does not mention this very important point.

'Old bees' is rather subjective and would be better defined as those bees that have completed their 3 week stint as house bees and have become foragers. These are the bees that are required for testing. Where is the best place to find them?

1. To be 100% sure then it must be those bees either leaving the hive or those returning and the sample can be taken from these but it is a bit tedious. Those leaving can be caught with a large polythene bag opened in front of the entrance and let 30 bees fly into it; count them on the way in. Alternatively, put a blanket over the entrance and wait until there are a sufficient number to scoop up in a match box. Both methods will ensure that only older bees are contained in the sample.
2. Young bees do not like the light and are likely to predominate in the centre of the brood nest. Old bees occupy the outer perimeter of the brood nest together with the returning foragers. Thus the first full frame of bees working from the side walls of the hive is the next best place to take the sample. This is used by most beekeepers who haven't the time to fiddle with collecting them outside. Taking them from the first available frame is quick and can be combined with a normal colony inspection. Note that care should be taken not to include the queen.

There is one place where the sample should not be taken and that is over the feed hole with an inverted honey jar. This method has been, and still is, recommended in various books and leaflets. The feed hole in a crown board is more often than not above the middle of the brood nest just where the young bees will be found and there is every likelihood of obtaining the wrong bees in the sample.

About 30 bees are required and a normal match box full holds about this number which can be checked by opening the match box over a piece of glass to see the contents.

2.11 Describe the factors that may initiate swarming and the indications that a colony is making preparations to swarm. Describe the economic and social effects of swarming and the procedures that are used to control swarming. Describe the procedures for creating an artificial swarm or any other method that may be used to ensure a colony does not swarm.

2.11.1 Describe the factors that may initiate swarming and the indications that a colony is making preparations to swarm.

2.11.1.1 General.

The MAFF booklet published in 1969 entitled 'Swarming of bees' is necessary reading for this part of the syllabus; unfortunately it is not in print and only old second hand copies are available. Its opening remarks liken a colony of honeybees to an immortal river whereby the dying bees are represented by the loss to the sea and the emerging bees to the tributaries maintaining the level. The life of the bee is finite and the colony is infinite which, of course, is untrue. Colonies perish for a variety of causes and they must reproduce; swarming is the way this is achieved to enable the species to continue to exist.

2.11.1.2 The origin of swarming.

Researchers have recognised that there have been (are?) two types of swarm as follows:

• Mating swarms whereby the colony or part of it accompany the queen on a mating flight and, presumably, never return to the original nest.
• Hunger swarms whereby the whole colony departs with the queen in times of dearth to find better foraging areas.

There are pros and cons for both theories being the origin of swarming of the subfamily of *Apinae* which contains *Apis, Melipona* and *Trigona*. The consensus of expert opinion is that the hunger swarm appears to be the most credible but a more detailed knowledge of the African bee and the stingless bees of S.America is required.

It is interesting to note that if the *Apis mellifera* from the temperate regions of Europe is compared with its present day ancestor from Africa there are some important differences related to swarming.

a) The European bee invests much of its colony resources in collecting and storing large quantities of honey and has developed the ability to cluster for long periods in order to survive a winter climate.
b) Survival in the tropics is considerably easier than in a temperate climate. The African bee invests much of its resources in swarms and drones and still swarms en masse to find forage,

often travelling large distances to find it and clustering overnight en route. This swarming instinct has been calculated as ×16 > than that of the European bee. The African bee does not cluster for long periods and does not have the same thermo-regulatory ability which, of course, is unnecessary in a warm climate.

c) The African bee has a preference for a nesting cavity of about 10 litres capacity whereas its European counterpart prefers 20 to 80 litres.

d) Its last major difference is its defensive trait, the African bee guarding and stinging at distances > 100 metres from its nest. For example, we have noted, while in Zambia, that the natives are very cautious when a swarm of bees is seen to be on the wing which happens very often after the jacaranda trees in that area stop blooming.

It will now be clear why the Africanised bee has spread so rapidly in South America and has caused such alarm because of its defensive traits. The propensity to swarm is part of the genetic make up of the African bee and this characteristic may be found in the Heath bees of Holland which have evolved from their African ancestors many millions of years ago.

2.11.1.3 Swarming theories and other causes.

After Langstroth discovered the concept of bee space in 1851 and the moveable frame hive came into widespread use throughout the world, practical beekeepers have endeavoured to minimise the swarming in their colonies in order to maximise their honey yields. Various theories have been advanced and the three which have received the most attention are as follows:

1. **Gerstung** (The brood food theory). This theory propounded by Gerstung of Germany in 1890 assumed that when a colony built up and became opulent there was an excess of brood food produced by an excess of nurse bees. So they built queen cells to use up this excess and swarming occurred.

2. **Demuth** (The overcrowding theory). Demuth, a very famous American beekeeper postulated this theory in 1921 and it received wide acclaim. Relieving congestion in the hive minimised swarming and this was extended by Demaree to congestion in the brood nest and hence his system of control to relieve this condition. From a practical beekeeping point of view the congestion theory was hailed as a great breakthrough in the management of colonies for honey production.

3. **Butler** (Queen substance theory). The discovery of 'queen substance' was made by Dr. Colin Butler at Rothamsted in the 1950s probably 1953. It simply postulates that pheromones are produced by the queen and required by the worker bees to inhibit the development of worker ovaries and to inhibit the building of queen cells in the colony. His experiments proved both these hypotheses and are extremely well documented in his book 'The World of the Honeybee' which also contains descriptions of some of the more important experiments that he undertook to prove the theory.

Butler's theory put paid to both the previous theories, in fact it explained the overcrowding theory. Queen substance is passed around the hive to each worker by reciprocal feeding one with another and any congestion would interrupt this process to a greater or lesser degree.

There are other factors which are attributed to swarming and said to be a trigger in the process. These are listed below:

• **Season.** There seems to be little doubt, by general observation, that swarming is more prevalent during some years when compared with others.
• **Weather.** Similarly, some weather patterns appear to be associated with the issue of swarms.
• **State of the flow** is said to affect the propensity to swarm.
• **Shade/ventilation** is also said to be a contributory factor.
• **Strain of bee.** There is little doubt about this factor; some strains are inveterate swarmers while some hardly ever swarm.
• **District or area.** It was shown that, during one year with a reasonably settled weather pattern over southern England, more swarming occurred in one county compared with another where statistics were being recorded. This was a one off sample and the sample is a small one and not too much credence can be given to it.
• **Manipulations** are said by some to trigger swarming. The beekeeper who leaves his colonies alone will never be certain of the extent of swarming and the authors are not convinced about this one.
• **Comb space.** This is directly related to congestion and is a contributory factor.

2.11.1.4 Queen substance (QS).

It is now known that the pheromone 'queen substance' is a complex mixture (31 found) of mainly fatty acids of which 13 have been identified. It is produced by the mandibular glands of the queen and also from other dermal glands on her abdomen. It has not been possible to synthesise it successfully to prevent swarming although it was attempted by Glaxo some years ago. The following points in relation to the pheromone are important when considering the swarming behaviour of the colony:

• Butler showed that the behaviour pattern of the colony depended on the actual amount the colony received, ie. it is a quantitative problem.
• J.B.Free has stated that a young (new) queen produces about 5000μg/day.
• A minimum threshold amount is required by each worker bee to prevent the building of queen cells. It is unclear whether the threshold amount is the same for all races and strains of bee. Similarly, it is unclear whether different queens produce the same amount per day. Compare the large prolific yellow colony with the smaller less prolific black colony. Does the yellow queen produce more QS or do the yellow workers require a lower threshold? Conversely, does the black colony require a higher threshold or does the black queen produce less?
• QS produced by the queen decreases with time and obeys the exponential law of decay. If she produces an average of 5000μg/day during her first year then this will halve (2500μg/day) during the second year and halve again (1250μg/day) in the third year, etc. The importance of maintaining a young queen to prevent swarming will be obvious.

2.11.1.5 The start of swarming in the colony.

When the supply of queen substance is below the threshold required for colony cohesion, the queen's egg laying rate will rapidly decrease because of reduced feeding of the queen by the workers. Those eggs which have been laid in the queen cups, which are a part of every normal colony, will not be removed but will be allowed to hatch out into larvae. Queen cells will result and the colony will be on its way to swarming. Other signs will be apparent as follows:

a) House bees will be reluctant to accept nectar loads from foragers.
b) Foraging diminishes and redundant foragers start to seek a new nesting site.
c) The queen ceases to be fed and decreases in weight by c. 30% to enable her to fly.
d) Egg laying virtually stops.
e) The decrease in foraging and brood rearing results in physiological changes to the worker bees. Because of the reduced level of QS, worker ovaries start to develop and also because of lack of brood rearing the hypopharyngeal glands also develop, producing additional fat bodies and an additional protein reserve.

The number of queen cells appears to depend very much on the strain of bee; those strains prone to swarming and genetically inclined that way will build very large numbers. We have counted as many as 70 in such colonies. For the more normal colony it is likely to be between 10 and 20.

By the time the first queen cell is sealed there will be no freshly laid eggs in the colony and the queen will be physically able to fly. Assuming the weather to be favourable the swarm will emerge at about noon. The following events are pertinent:

a) The emergence is preceded by the 'whir' or 'buzz' dance where the bees run backwards and forwards across the combs in horizontal lines buzzing with half spread wings every 0.5 to 3.0 seconds. Buzz frequency is c. 180 to 250 cps.
b) The dancers touch other bees (for up to 5 seconds) buzzing at 400 to 500 cps when touching.
c) The disturbance and excitement multiply and soon lead to the emergence of the swarm. Exactly which bees go is unknown and some return very quickly to the hive.
d) We have not been able to trace at what stage the bees gorge themselves with honey before departure from the hive.
e) The prime swarm with the old queen will settle usually within 10 to 20 metres of the hive.
f) Queen pheromone and Nasonov pheromone are vital to co-ordinate the swarm while in flight and during the settling process.

2.11.2 Describe the economic and social effects of swarming.

The syllabus does not make it clear whether the economic effects of swarming on the parent colony is required or the economics with respect to the swarm itself; both are of considerable interest. Similarly, it is not clear whether the social effects on the parent colony resulting from the issue of a prime swarm are required or the social effects of the swarm establishing itself and developing to a state whereby it can survive the forthcoming winter. All 4 options should be briefly examined.

2.11.2 1 The economic effects of swarming.

a) With respect to the parent colony:
• After the prime swarm has issued the colony is reduced in strength by c. 50%.
• The foraging force is drastically reduced.
• The honey crop is also, in turn, drastically reduced.
• The possibility exists that the parent will throw one or more casts and become an uneconomic unit.
• If this happens it may not survive the coming winter.

b) With respect to the swarm itself:
• The swarm can carry only meagre resources. The average bee in a swarm carries 35mg of honey with a 65% sugar content. Thus a swarm of 12,000 bees carry 273g sugar. They carry no pollen or water.
• This reserve is sufficient to build only 1,100 cm^2 of comb.
• It needs to start foraging immediately to survive by building more comb for stores and brood rearing.
• If the weather is inclement the swarm is likely to perish.

2.11.2.2 The social effects of swarming.

a) With respect to the parent colony:
• The life of the colony is in abeyance until a new fertile queen has been established.
• The colony will be queenless for c. 8 days after the swarm has departed.
• A further 5/6 days minimum will elapse before a queen is mated and starts to lay.
• The queen can be lost due to predators while on mating flights.
• Further swarms can emerge further reducing the strength of the original colony.
• During the period of queenlessness the colony is not functioning as a normal social unit.

b) With respect to the swarm itself:
• The process of determining a new nest site (a cavity of 20 to 80 litres).
• The process of initiating comb building and then building sufficient comb in the chosen cavity.
• No brood rearing can commence without comb so this is of very high priority.
• As the queen is old then supersedure is most likely to occur before the drones are evicted.

2.11.2.3 We have touched only very briefly on the 4 topics above; all are sufficiently complex to run to many pages of detailed explanation. Our reason for not going into greater depth is that the topic appears to be a small part of the overall syllabus dealing with swarming and all four topics are to do mainly with honeybee behaviour and this is essentially a practical examination.

2.11.3 Describe the procedures that are used to control swarming.

Before describing the procedures used to control swarming it is essential to endeavour to prevent swarming, then to be able to detect swarming preparations involving inspections at known intervals and then finally comes the actual control.

2.11.3.1 The definitions and what is involved:

• **Swarm prevention:** is the action(s) taken by the beekeeper to prevent the colony reaching the state whereby it starts to build queen cells.
• **Detection of swarming preparations:** this is necessary before any swarm control measures are put into practice. The ability to detect the preparations is prerequisite to any control actions.
• **Frequency of inspections:** for swarm detection is dependent on whether the queen is clipped or unclipped.
• **Swarm control:** is the action(s) taken by the beekeeper to thwart the colony in its endeavours to swarm once the preparations for swarming have been started thereby preventing the loss of bees.

Each will be examined separately. However, a further basic distinction must be made and that is between the stocks in the home (or fixed) apiary and those stocks which have been moved for pollination or to exploit a source of nectar (eg. rape in the spring). The control method is likely to vary depending on whether the stock is close to hand where additional spare equipment is readily available or on a remote site where spare equipment is not readily available. On this basis two methods of control will be discussed.

It is necessary to control swarming for a number of reasons which are frequently overlooked by many beekeepers, these are:

• A colony that swarms is unlikely to produce a surplus cf. the colony that does not lose its bees; this is to the detriment of the beekeeper but of little consequence to anyone else.
• Most of the general public are petrified of bees and if not petrified then they have an innate 'api-phobia'. In an urban environment it is essential that no swarm settles on a neighbouring property (this cannot be guaranteed).
• When a colony swarms, there are many thousands of bees flying around which most people find very frightening and can be classed as a nuisance in an urban or suburban environment.

2.11.3.2 Swarm prevention.

It is imperative that the role of queen substance in relation to swarming is thoroughly understood before proceeding further. It is important to understand the role of queen substance and the inter-relationship between food sharing and congestion in the colony as the trigger in the process of swarming. The prerequisite in swarm prevention is that the colony must be headed by a young queen in order that each bee in the colony is assured of its minimum threshold quota of queen substance. When an adequate supply is available at its source (ie. the queen), the next most important factor in swarm prevention is to ensure that the supply can be distributed around the colony; this can only happen if there is plenty of comb for the bees and hence no congestion. This means the timely addition of supers as the colony is expanding. Add good hive ventilation and the beekeeper can do little else in the way of prevention. Nevertheless, having done all this a colony may proceed to build queen cells and it is the beekeeper's job to control the issue of a swarm.

2.11.3.3 Detection of swarming preparations.

It is very important for every beekeeper to be able to recognise the preparations for swarming while undertaking a routine inspection. At the beginning of the season the colony will have no drones and no queen cups (easily recognised; being almost identical in shape and size to acorn cups). As the colony builds up drones will appear and queen cups (known in some parts of the country as play cells; reason unknown) will be built around the outer limits of the brood nest. It is important to examine them closely. If eggs are found in them it does not follow they will be turned into queen cells; in many cases the eggs are eaten by the bees. However, if the cup contains royal jelly, a larva will also be present which is sometimes difficult to see floating in the pool of liquid as the egg may have only recently hatched. This is the sign that preparations for swarming have commenced and swarm control proceedings must be initiated. The simple rules are:

1. Dry queen cups (nothing in them or egg only); the situation can be left to the next inspection.
2. Charged queen cups (containing royal jelly); initiate swarm control procedures.

If all the queen cups have a dull matt finish on the inside, preparations for swarming have definitely not started; the cells will be polished before the queen will lay in them.

When swarming is imminent there will be a marked reduction in the laying of the queen and fewer eggs are likely to be seen.

2.11.3.4 The frequency of inspections.

It is important that every colony is inspected, regularly, throughout the active season. The frequency of these inspections depends on whether the queen is clipped or unclipped. Briefly the rules are as follows:

Queen unclipped - inspection every 7 days without fail.
Queen clipped - inspection every 14 days until preparations for swarming are detected and then every 10 days without fail.

For a complete understanding of how these time periods are determined reference should be made to the Green Book, Beekeeping Study Notes for Module 1 Section 1.17 and the associated Appendix 4 - Colony Inspections - timing.

2.11.3.5 Swarm control.

The Queen Substance theory (postulated and **proved** by Butler in 1953) is the only theory that satisfactorily explains why a colony swarms and is now accepted as the only correct theory of swarming. Congestion prevents queen substance from being distributed around the colony and is therefore, in itself, not a theory. The brood food theory was accepted for a long time but is now regarded as being incorrect; it is based on the surmise that as the colony builds up, an excess of brood food is produced and this is used in queen cells that are built to absorb this surplus.

Most swarm control methods involve finding the queen and some require finding and destroying queen cells which in turn requires shaking bees off frames. Allied with these operations of controlling, regular inspections are required to know when to undertake them. Such inspections and control can only be undertaken with good tempered bees and ensuring the 'right strain' is **a necessary part** of swarm control. When the colony becomes bad tempered, regular inspections get abandoned, the colony swarms and the bad temper is promulgated further around the district. This indeed must be classed as anti-social behaviour on the part of the beekeeper. The Authors believe that a major contributory cause of such situations arising is the present day obsession to wear gloves to manipulate the colony. If the norm were no gloves (kept in reserve for the real emergency) then colonies would be requeened before situations got out of hand. If colonies cannot be handled without gloves, then the handling technique or the strain of bee is at fault and should be corrected without delay. Anyone keeping bees in an urban garden should consider this point long and hard.

There are innumerable methods of swarm control from the destruction of queen cells, Demareeing, Snelgrove system, Pagden/Artificial swarm, etc., etc. Only one of these, the artificial swarm is required for this syllabus and is described in detail below.

2.11.4 Describe the procedures for creating an artificial swarm or any other method that may be used to ensure a colony does not swarm.

The Artificial Swarm. This method, which must be common knowledge to anyone keeping bees, is so well known and documented that only a few comments are necessary. Briefly, when the operation has been completed the queen and one frame of bees plus empty comb to fill a new brood chamber remain on the original site and the colony with all the queen cells and remaining bees is put on a new site within the same apiary. All foraging bees return to the original site and, with the queen, form the artificial swarm. The old colony with only house bees and queen cells rear a new queen without swarming. This is the basis of the method, other points of interest are as follows:

a) If the colony has supers, then where should these end up; on the artificial swarm with the foragers or on the original stock with the queen cells? Most books show the supers on top of the artificial swarm on the original site. The old stock (now weakened by c.$\frac{1}{3}$ of the total original number of bees) on a new site may need feeding and could be robbed. It seems logical to put the supers on the old stock and feed the artificial swarm which in all probability will have foundation to pull out and also, doing it this way, there will be no possibility of contaminating the supers with sugar syrup.

b) Again many books recommend moving the original stock a second time to draw off any additional foragers 7 days after the manipulation and before a virgin has emerged. Unfortunately the rationale behind such a move is not explained. It does provide additional foragers for the artificial swarm but is not essential to the success of the manipulation.

c) It is unnecessary to destroy all but one queen cell in the original stock as the removal of foragers reduces drastically the strength of the colony and the bees will undertake the destruction themselves.

d) If necessary, the operation can be completed on the same site with the artificial swarm below and the old stock on top above a swarm board or similar. If it is done this way, then any feeding will be confined normally to the top stock.

e) The advantage of this method is that it is virtually 100% successful and can be performed on any stock. Additionally, brood rearing continues with the old queen and the two units can be united at a time suitable to the beekeeper. The disadvantage is that additional equipment is required. As an example, the authors had 8 colonies on the rape in 1988 and all wanted to swarm; to use the artificial swarm method was just not practicable away from our home apiary, and further we did not have enough equipment available at the time. There are horses for courses, and the beekeeper has to make up his own mind how to manage the situation and which method of swarm control to use.

f) This method of swarm control requires the beekeeper to find the queen. It is infinitely easier to find the queen if she is marked and we regard marking as high priority for effective swarm control.

2.12 Describe the procedures used up to the time of the assessment in the queen rearing method demonstrated and explain what has yet to be done. Describe what is intended for the queens that have successfully mated. Describe the procedure that will be adopted to introduce queens into a colony.

2.12.1 Queen rearing method. These techniques have been described in Section 2.6. The candidate should organise his queen rearing programme to coincide with the pre-arranged date of the visit from the assessor.
The candidate should be able to demonstrate to the assessor the procedure he is using to rear queens in his apiary, and by producing a frame from the cell building colony which shows acceptance of introduced one day old larvae or sealed queen cells, show his competence with the method chosen.

2.12.2 Uses of the mated queens.

We are unsure of the intent of this part of the syllabus because any queens that are reared are more than likely to be used to requeen colonies with ageing queens. There can be one exception to this; some may be culled on the grounds of producing aggressive offspring. Our own method is to rear the queens in May and keep them in 5 frame (BS frames) nucs until next February/March when they are introduced into existing colonies; the old queens are put into the nucs to keep them going until further new queens are reared again in May. During the time from May to March the temper of the queens offspring can be assessed and only those that are well disposed are used in full colonies.

2.12.3 Introduction of a mated queen into a colony.

2.12.3.1 General considerations.

The ability to requeen a colony of honeybees with a young fertile queen has been described as the pivotal point of beekeeping. It enables, if successful, the colony to be rejuvenated and if the queen produces 'good' progeny, the future of the colony is ensured. Practically every writer seems to discuss the subject, and often pontificate, but the fact remains that every year many fine queens are lost through unsatisfactory introduction methods. This suggests that either the methods being used are unsatisfactory or the beekeepers are doing something wrong. Re-queening successfully is a feature of good beekeeping; conversely, re-queening unsuccessfully must be the mark of bad beekeeping.

Laidlaw probably summed it up correctly when he said that no method is perfect, some are more successful than others. We believe that we want to be quite precise about this topic, which is the introduction of fertile queens or mated queens, although we will address later, for completeness, the introduction of virgins under certain conditions. We must be precise also about what we are introducing the queens to and that the final objective here must be a full sized colony.

2.12.3.2 The principles underlying introduction.

The principle involved is extremely simple. The colony must be in a condition to receive a new queen and that queen must fulfill immediately the requirements of the colony she is introduced to. The problems lie in fulfilling the conditions on both sides of the equation, the colony on one side and the queen on the other.

From the colony point of view it must be queenless and from the queen point of view she must replace what has been removed. Bro. Adam maintained that it is not an introduction but a substitution.

2.12.3.3 Methods and precautions to be taken.

The methods can be addressed as indirect and direct introduction, ie. by introduction cage or without one.

Bro. Adam always maintained that the successful introduction of a queen was solely dependent on her behaviour and that she had to behave sedately and be in lay at the time she was introduced or released into the new colony. He further stated that a young newly mated queen is capricious and nervous, taking from 4 to 8 weeks to change her behaviour pattern whilst becoming in that time sedate and quiet on the comb, going about her business of laying eggs. A queen of this disposition can be substituted into another colony with reliability. His findings were not confined to any one particular race but were true across the whole spectrum of races that he experimented with. He had one proviso that queens of the dark European races took the longest time to reach a stage where they could be introduced with any reliability and here he stated 8 weeks old as a minimum.

2.12.3.3.1 Indirect methods.

We have identified 30 to 40 different methods (or cages) all based on much the same principle of imprisoning the queen within the colony for a certain time before she is released either by the bees (eating their way through candy) or by the beekeeper. We propose discussing only one in detail as all the others are similar and can be found in the general literature. We know of only one book on the subject (although the subject seems to be addressed by most beekeeper writers) and that is by Snelgrove, 'The Introduction of Queen Bees', now out of print and first published in 1940 so it is a bit outdated.

The Butler Cage. This must be the most well known cage at the present time and was designed by Butler when he was undertaking his research on queen substance at Rothamsted. It is made of wire mesh (8 per inch) in the form of an oblong $\frac{1}{2} \times \frac{3}{4} \times 3$ inches long plugged permanently at one end with a wooden plug, the opposite end is left open. The size of the mesh is important as it is large enough to allow the bees outside to have contact with the queen and more importantly to feed her. Unfortunately few of the commercial suppliers can offer such a cage and most if not all use a mesh which is far too fine restricting the very important contact which is required. The $\frac{1}{2}$ inch (2 bee spaces) dimension is important to allow the cage to be placed in the colony between the frames in the brood chamber where the spacing between the comb faces is $\frac{1}{2}$ inch under normal circumstances.

A laying queen is taken from a nucleus or other colony and placed in the cage which is either plugged with a wooden plug or sealed with a small piece of newspaper over the open end and held down with a small elastic band. The colony to be requeened is dequeened and the new queen in the Butler Cage placed in the colony brood nest over or adjacent to some emerging brood. The queen is released the next day by the bees or the cage is removed the next day or at some other convenient time. The newspaper seal can be perforated with 2 or 4 pin pricks to start the release by the bees, but this is not essential.

Our bees, like those of most other beekeepers, are mongrels and we always prefer to supervise the release of the queen into one of our colonies generally after the queen has been confined for two days. When the colony is opened up and the queen is active in the cage, this immediately indicates that the house bees have been feeding her; a good sign. The wooden plug is removed and the queen allowed to walk out. Usually she starts her work immediately looking for a cell to lay and the bees start to groom her. If the activity is anything other than sedate she goes back into the cage for another 2 days and so on until she is accepted. We have had one occasion when it took 8 days and this was a yellow queen into some very dark and nervous bees. We believe that had we let the bees release the queen by the newspaper method she would have been lost. We advise, with confidence after many hundreds of introductions, that it is better to supervise the release rather than leave it to chance. We do about 30 to 40 per year and during the last 10 years we have only lost two queens; one we believe was our fault in carelessly handling the queen putting her in the cage and the other was a queen from New Zealand which was introduced to a small nucleus on arrival. Two days after this queen was introduced she was found dead in the cage in the nucleus; clearly the bees had not been feeding her for reasons best known to themselves.

Let us consider the precautions to be taken using the Butler Cage.

1. The colony (or nucleus) must be queenless. This may be obvious but we are certain that many queens are lost on this score alone. It amounts to carelessness on the part of the beekeeper.

2. There must be no trace of any queen cells either open or sealed in the colony. Most races of bee will prefer to raise their own queen at the expense of the one being introduced. Queen cells produce a pheromone which has the same effect as queen substance; if this was not so, laying workers would be produced in the case of an imperfect supersedure.

3. The time of introduction is important. We have found that the best time for requeening is in the spring (March) when the colonies are small. We have been told on many occasions that opening colonies in March will chill the brood; we disagree and confirm that we have never seen chilled brood in a colony. We can only reproduce the condition in the refrigerator for our winter evening classes. Most good practical beekeepers are aware that requeening during a honey flow is likely to be more reliable than when there is a dearth. Thus if at the time of introduction there is no flow on, it can be simulated by feeding; the feeding should be started prior to the day of introduction.

4. There should be no laying workers in the colony to be requeened. It is virtually impossible to requeen such a colony by this method. Experiments in France have shown a fairly high success rate by coating the queen with a mixture of royal jelly and distilled water. We believe this to be of academic interest only; such a colony is a useless unit when it reaches this state.

5. Generally, virgins and fertile queens that are not in lay should not be introduced by this method into a full colony. A travelled queen in a travelling cage will not be in lay and needs special treatment. It will be clear that fertile queens are fed only royal jelly and it is the attendant bees that provide this directly from their hypopharyngeal glands while the queen and attendants are in transit. If the workers or some of them are dead on arrival it does not speak highly of the queen breeder; he has selected bees of the wrong age to be the attendants. If the queen has not been fed well she will not be in the correct state for introduction.

6. There must be no robbing in progress and the colony must not be hungry; both these conditions are associated with bad beekeeping.

7. The size of the colony should be as small as possible although this condition is sometimes not controllable except by taking out a nucleus, introducing the new queen to the nucleus and later uniting the nucleus by the newspaper method with the parent colony. Here a word of warning. If a yellow queen has been introduced successfully to a nucleus made up of black bees, re-cage the queen in the nucleus during the uniting process in order to prevent her being balled. We believe that when a colony rejects a queen, she is not stung to death but balled and killed by the high temperatures in the ball. Spring is the best time for queen introduction and this is the time that the colony is at its smallest.

8. The colony must have some young bees in order to feed the queen. This is one of the reasons why laying workers are difficult to requeen, there are no bees producing brood food or royal jelly. A travelled queen in poor condition can be introduced successfully by making up a small nucleus (say 3 frames, one of emerging brood and two of pollen and liquid stores) with only one frame of nurse bees to cover the emerging brood and maintain the incubating process. The object is that all bees as far as possible shall be capable of feeding the queen initially through

the cage mesh. The entrance shall be one bee space and ideally the nuc should be isolated from another apiary.

9. It has been stated that premature examination after introduction leads to loss (balling, probably). We have not found this to be the general case.

10. Finally on precautions, strain of bee is important. There is an inherent difficulty introducing yellow queens (of Italian or New Zealand origin) to black bees but not the other way round. The reason for this is obscure but may have something to do with queen substance and the threshold amount required by each worker bee under normal circumstances. Consider the two colonies. The black bee is not so prolific and its colonies are smaller than the yellow bees which are more prolific. Do the workers in each require the same threshold of queen substance or do they have different threshold levels? Do the queens produce roughly the same amount or does the yellow queen produce more than the black? These two questions remain unanswered as far as we know.

11. The queen must be handled with great care at all times.

When a queen arrives by post or any other means it is important to remove the attendant workers just before she is introduced into the nucleus. The reason will be obvious from the notes above but we know of cases where they have been put into the hive in the travelling/introduction cage lock stock and barrel with the queen. Curiously we have not seen any reference in the literature to the queen being on her own in the introduction cage.

2.12.3.3.2 Direct methods.

Almost all direct methods of queen introduction are generally regarded as gimmicks in these days of modern beekeeping; however, there are one or two occasions when they may be used.

There are a variety of methods claimed for direct introduction from shaking the dequeened colony out in front of the hive (shook swarm) and, in the resulting confusion, running in the new queen with the bees to dunking the queen in a variety of substances (eg. water, honey, etc.). None are now recognised as modern methods and generally are considered a bit cranky even though some have been advocated by eminent beekeepers (eg. Simmins). After dunking, the advocates then run the queen in via the entrance or the feed hole at the top.

The major problem with running queens into a colony is the lack of control being exercised by the beekeeper and not knowing whether the introduction has been successful unless the queen is marked.

2.12.3.4 Attendant difficulties; the strains of bee and colony conditions have been discussed above under indirect introduction.

2.12.3.5 Introducing virgin queens.

It is notoriously difficult to introduce virgin queens to either a small nucleus or a larger colony. That such a colony must be queenless, it goes with out saying, and in general must be absolutely and hopelessly queenless to achieve any success at all. The reason is very simple. A virgin queen produces virtually no queen substance when she is newly emerged and has little attraction for the workers. It is only as she matures during her first 5 days after emerging from the queen cell that the amount of queen substance increases up to the stage where she is capable of attracting drones for mating. It will be clear that the best chances of success are with virgins about 4 days old and they should be introduced within 24 hours so that, given the correct weather, they may go out to mate on their 5th or 6th day which would be their normal behaviour pattern. The smaller the nucleus the better made up with young bees whose hypopharyngeal glands are in the right condition for feeding the queen. There is a possibility of the whole nucleus absconding with the queen as a mating swarm. We still continue to experiment with the introduction of virgins to mini nucs and have evolved a fairly reliable method.

2.12.3.6 Other points of interest.

During 1989 there was some serious work undertaken in France on direct introduction by bathing the queen in a mixture of royal jelly and water (70% & 30% respectively). Trials were carried out on both fertile queens and virgins into nucs, normal colonies and colonies of laying workers. The success rate was high for all permutations and it is expected that more will be heard of this approach in the future.

The reader should be alert to the fact that queenless colonies attract queens from swarms. Also that well stocked apiaries attract stray swarms from outside sources. One year we had our cell builder ruined by a stray queen(s) from a stray swarm that settled in the apiary.

2.13 Demonstrate marking and clipping a queen, or use a drone as a substitute if appropriate.

There is little we can say on this part of the syllabus, you the Candidate are on your own 'flying solo' but do be alert to the fact that, although you have practised on drones, an examiner may well ask you to find and mark the queen in one of your colonies. The moral here is to have all your colonies marked and clipped for the examination! See Section 2.14 blow.

2.14 Describe the advantages of marking and clipping queens.

In order to mark or clip a queen it is necessary to keep her still and in such a position that the operation may be effected without damaging her. It is an aid to bee management but our own observations reveal that marking and clipping are not widely practised by hobbyist beekeepers in the UK. Some beekeepers object to clipping on ethical grounds, a point of view we cannot understand and seemingly those who hold such views cannot explain them.

2.14.1 Queen marking cages.

There are two basic types of queen marking cages, namely the press on cage which imprisons the queen in the cage on the comb and the glass plunger type whereby the queen is gently pushed to the end of the tube with a soft foam plunger until she is just in contact with a plastic grill through which she is marked.

There are two types of press-on cage one made of plastic and the other called a Baldock type which is prefabricated. Both are about $1^1/_2$ in diameter with a mesh on top having about $^5/_{32}$ in squares. The bottom has a series of spikes around the bottom to push into the wax comb. The only difference between them is the Baldock type has finer wires forming the top mesh making it easier to mark the queen and it has finer pointed spikes making it easier to push into the comb. We recommend the Baldock type, the plastic one is somewhat crude and more difficult to use.

The plunger type requires the queen to be handled. She has to be put into the glass tube at one end and pushed gently up to the other end. It unfortunately has a plastic grill and suffers from the same defect as the plastic push-on cage, the plastic mesh of the grill is too coarse. We have never used one but have seen queens damaged by clumsy operator techniques.

2.14.2 Holding the queen in the fingers.

Many beekeepers, including ourselves, never use cages for marking. We find it easier to pick the queen up and hold her while clipping and marking. The recommended way is to pick the queen up by her wings in the right hand (assuming you are right handed) and hold her between the thumb and forefinger while preventing her slipping down with the second finger beneath the queen. She is held by the head and thorax not the abdomen.

We use a different method which was shown to us many years ago by Brian Palmer of Kent. We have used it ever since and we have shown many beekeepers how to do it; to our knowledge we have never seen mention of it in print. The queen is picked up in the same way by the wings and placed on the tip of the first finger of the left hand. The three legs on one side are smoothed down with the thumb, also of the left hand, until all three are trapped between the thumb and fore finger. When she is safely held by the legs the hold on the wings may be released. It is surprising the amount of pressure that can be used to hold the three legs without damaging the queen. We suggest that practice is first gained with a few drones.

2.14.3 Marking queens.

2.14.3.1 Reasons for marking.

These are the advantages of marking and are listed in order of popularity:

> a) To make the finding of the queen easier.
> b) To identify the age of the queen.
> c) To identify different suppliers and strains.
> d) Research and studying queen behaviour.
> e) For use in observation hives.

2.14.3.2 Materials used for marking queens.

Clearly the colour is important particularly with dark bees where the queens are much more

73

difficult to find compared with the lighter coloured bees. White and yellow are particularly good. Most bee suppliers can provide a set of colours in quick drying paints or enamels which are normally used for model aeroplanes or motor cars. It is important that emphasis is put on the quick drying and here the solvent is important. Any use of amyl acetate (which smells of pear drops) should be avoided as it has similar effects to isopentyl acetate (which smells of bananas) the alarm pheromone from the sting chamber. Most nail varnishes fall into this category. We do not colour code our queens and we always use white Tipp-Ex (as used by stone age typists) which uses trichloroethylene, an industrial degreasing agent. It is to be noted that Tipp-Ex is now marked on the container stating that the solvents are aliphatic hydrocarbons and not trichloroethylene which take longer to dry and may cause balling of the queen.

If colour coding is used, indicating the year the queen emerged, then there is an internationally agreed colour code as follows:

COLOUR	LAST DIGIT OF YEAR	READ DOWN
White	1 or 6	Which or When
Yellow	2 or 7	Year or You
Red	3 or 8	Reared or Requeen
Green	4 or 9	Great or Give the
Blue	5 or 0	Britain or Best

Paints, enamels or other marking fluids can be applied in a number of ways with a small brush (Tipp-Ex comes complete with brush in the lid), head of a match stick, pin stuck in a cork, etc. With all marking it is important that the marker is sufficiently fluid to penetrate the small hairs on the queens thorax and adhere to the exoskeleton of the thorax, the only place a queen should be marked. The amount used must be just right and must not run down into the petiole or into the queen's eyes.

Small numbered discs are available which are attached to the queen with a small dab of quick drying glue. These are also available in the colours quoted above.

Another variation is the Eckhart marker which cuts out a small circular disc and is ejected and stuck onto the queen with glue by an internal plunger. They are not often seen these days but were the 'in' gadget 50 years ago.

Fluorescent paint has been used in conjunction with UV light and radio active paint in conjunction with a Geiger counter. It is unlikely that the hobbyist will have any interest in these two methods which are much more research orientated.

It is important that the queen is held for a short time after marking for the paint to dry or the glue to set before she is released back into the colony.

2.14.3.3 The disadvantages of marking.

There are a few disadvantages in marking which are as follows:

a) Whenever the queen is handled or touched there is always a small danger that she may be damaged. The moral is handle her very carefully.

b) Marking with the wrong type of marking material can prove fatal (nb. amylacetate).

c) Badly marked with an excess of marking fluid will prompt the bees to supersede the queen.

d) One source we consulted indicated that marking a queen leads to supersedure. We have marked hundreds and this has not been our experience.

2.14.3.4 When to mark a queen.

It is our belief that all queens should be marked. There are two reasons for saying this:

1. Practically all methods of swarm control require the beekeeper to find the queen and a well marked queen makes the task so much easier and more reliable particularly if the bees are a bit 'sharp'.

2. If a colony becomes bad tempered and the queen has to be changed. To do this quickly and efficiently great concentration is required to find an unmarked queen particularly if the bees are also runners.

There is only one time to mark the queen, in our opinion, and that is when the colony is small and easy to handle. This means in the spring just after the first inspection. We never touch or handle a queen from the time she emerges (usually about May) until about March when she is both marked and clipped before introduction to a colony from an over wintered nucleus. Bro.Adam advocated never touching a queen until she was at least 8 weeks old; very sound advice. The practice of marking virgin queens is not recommended.

2.14.4 Clipping queens.

2.14.4.1 The reasons for clipping queens.

The primary advantage is in connection with swarming. Using a clipped queen it is possible to have 14 day inspections up to the time swarming preparations are detected and then the beekeeper must revert to 9 or 10 day inspections. If a swarm issues with a clipped queen the queen falls to the ground and the swarm of worker bees returns to the hive. Using the aforesaid inspection times saves losing the bees which are the work force for collecting a crop of honey. The secondary advantage is to provide an indication of the age of the queen by clipping a particular wing for odd years and the other for even years. This is a form of marking.

2.14.4.2 Clipping.

About $1/3$ of the wing (s) is cut off. Single wings or both on one side or both on both sides may be trimmed. The important point about clipping is to have a very sharp pair of scissors with very fine points. Surgical scissors are by far the best.

Be very careful about the queen curling one of her legs over the top of her abdomen when the clipping operation is being done. It is very difficult (nearly impossible in our opinion) to clip a queen in a queen marking cage. Beekeepers try it and we suspect that the queens are damaged in the process. We believe that to clip a queen it is necessary to hold her in one hand and clip with the other.

It is very important that one's fingers are not sticky with propolis when picking up a queen by her wings; if they are really sticky it will be very difficult to release her without damaging her at the wing roots or her legs. This is the only time gloves are permissible; wear them until the queen has been found and then remove them to do the clipping. At these times we adopt a single role, one looking for the queen and the other doing the marking and clipping with a clean pair of hands. A much better arrangement.

2.14.4.3 Other points.

• Clipping does not hurt a queen, it is rather like having a hair cut. We have met some beekeepers who will not clip their queens and say it is wrong to do so but the reasons why it is wrong are never forthcoming.
• Often clipped queens walk back into the hive after swarming. As long as one is alert to this it should be no surprise to find a clipped queen and sealed queen cells in the colony.
• Sometimes if the queen does not get into the hive she will be found with a swarm under the hive stand depending of course on whether she can gain access to the underside.
• Once a queen has been clipped it is difficult to pick her up if she has to be transferred to another place. In such cases we always use a Butler cage put over the queen on the comb and wait until she walks up into the cage before plugging it with the queen inside.
• If the beekeeper's eyesight is defective then spectacles are essential for this delicate operation.

2.15 Describe the procedures adopted for adding supers.

The principles of supering are, generally, inadequately dealt with in the classical literature; supers are basically for bees not for honey. If the supers are filled with honey then this should be regarded as a bonus for good management. Because they are for bees, supering is the primary method of preventing swarming.

2.15.1 The principles of supering.

2.15.1.1 Definitions:

> **Super**. A box containing frames/combs placed above the brood chamber for the eventual storage of honey. The word 'super' is derived from the Latin word super meaning above (eg. super-script as opposed to sub-script). Supers are generally shallower than brood chambers because of the weight when full of honey; other than this, there is no technical reason why they shouldn't be any depth providing the frames can be accommodated in the extractor.

> **Supering:** is the process of adding supers to a colony above the brood chamber either with or

without a queen excluder under the super(s).

Top supering: is the term given to adding further supers to a colony but always adding them on top of any existing supers.

Bottom supering: no prizes for guessing that the supers are added at the bottom of the pile and always next to the brood chamber.

2.15.1.2 Principles involved.

Reference to the annual colony population cycle graph, Appendix 8, shows the very rapid increase in adult bee population from the beginning of March. It is not long, providing the weather is fine, that the brood chamber starts to fill up with both brood and bees and if nothing is done there will be insufficient room for the emerging brood. Additional space is therefore provided by adding supers, usually one at a time, as required by the colony build up. On this basis, supers are for bees and, indeed, this can be very true if the colony is using most of or all its income. In such a situation nothing will be stored in the supers and it will be used solely as a parking place for bees in the colony. If this additional space is not provided, overcrowding will occur and this congestion in the hive leads to a breakdown in the food sharing pattern and subsequent distribution of queen substance with a result that the liability to swarm is greatly enhanced.

When the honeybee undertakes the manipulation and ripening of nectar to honey, large areas of comb are required for the nectar/honey to be 'hung up' to dry in order to evaporate the water. The change in volume of nectar (30% sugar concentration by weight) to honey (80% sugar by weight) is approximately 100:30 thus requiring c. 3.3 times the space for nectar compared with the space required by the finished product.

The calculation looks like this:

> 1 litre honey weighs c. 1400g   (80% sugar, 20% water by weight)
> 1 litre of water weighs 1000g

and
> 1400g honey = 1120g sugar + 280g water
> 1000ml honey = 720ml sugar + 280ml water
> 1g sugar has a volume of $720 \div 1120 = 0.64$ml

CONCENTRATION	SUGAR	WATER	TOTAL
Nectar 30%	30g	70g	100g
	19.2ml	70ml	89.2ml
Honey 80%	30g	7.5g	37.5g
	19.2ml	7.5ml	26.7ml

It will be seen that 89.2ml of nectar (30%) when processed to honey only requires a volume of 26.7ml; this is a change of $89.2 \div 26.7 = 3.3$.

There are only two principles involved as detailed above and summarised below:

1. Primarily to provide space for bees, and
2. To provide comb area for ripening nectar.

If adequate space is provided for evaporation then it will be clear that there will be adequate space for honey storage.

2.15.1.3 Other points related to supering:

a) By experience it has been found that a good working guide for supering is to add a super when the bees are covering all but the two outside frames of the top box or initially the brood chamber.

b) It is better to super early in the spring and be somewhat tardy about adding supers in July when the main flow is on unless this is absolutely necessary.

c) In general, top supering is the most widely used method of adding supers. Bottom supering is advantageous if the frames in the super contain only foundation; ie. placing them above the brood chamber, the warmest place in the hive for the wax makers to work.

d) There are quite a few beekeepers that super without the use of a queen excluder; however, the majority use an excluder. Again there are beekeepers who advocate not using an excluder in the spring, when the first super goes on, to encourage the bees into it more quickly. It is true the bees always seem to be somewhat tardy about occupying the first super but this may be due to observing the rule of being just a little ahead of the bees' requirements (super when the two outside frames are uncovered).

e) See Section 1.5 of Beekeeping Study Notes Modules 1 to 4, the Green Book, on spacing of frames. Narrow spacing is essential when starting with foundation which can be widened out to 2in (51mm) when drawn and being filled with honey. The maximum of 2in (51mm) is the maximum that a colony will build for the storage of honey in the wild state and cut comb containers have been designed on this thickness.

f) If the super contains frames with foundation only, one or two frames of drawn comb in the middle will encourage the bees into the super more quickly.

g) Wet stored supers are more attractive to the bees in the spring cf. dry supers.

h) The first super above the queen excluder is likely to have some pollen filled cells in it. Pollen will always be stored above the brood where it is to be used. Ensure that supers for cut comb or any kind of comb honey are above the first super. Pollen in comb honey is unacceptable for sale; it has a bitter taste.

2.16 Describe the procedures adopted when removing supers for honey extraction.

When removing supers for honey extraction it is necessary to clear them of bees.

2.16.1 General points on clearing bees.

• By definition, clearing implies a crop has been collected and the flow is over; robbing can easily be started unless care is taken when removing the crop.
• Bees are generally more irritable after the flow and will be more inclined to defend their stores than before the flow finished.
• Entrances must be reduced at the same time as supers are being cleared.
• If more than one super has been used it is common for brace comb to have been built joining the supers together, the brace comb being filled with honey. It is essential to remove this brace comb from the top and bottom bars of the frames 24 hours before clearing, in order to avoid honey dripping from the supers as they are removed. It is a sticky job to do but well worth having the frames cleaned up and no honey dripping while they are being collected. The brace comb should not be there and emphasises the importance of bee space and the many incorrect frames that are in use. The process known as 'cracking the supers' is not given the attention it needs in modern bee literature. It should be done just before dark. It also prevents the surprise of finding the supers not cleared because of brood in the one next to the brood chamber, if the first supers are checked during the cracking process.
• Supers should be removed very early in the morning before the colony has started flying and taken straight to the extracting room for extraction the same day.

2.16.2 Clearer boards.

There are basically two types one using Porter Bee Escapes and the other called the Canadian type with long tunnels for the bees to traverse to get from one side of the board to the other.

Porter bee escapes: possibly the most popular device in UK for clearing bees. The following are the salient points about its use:

a) The phosphor bronze springs require very delicate adjustment to a gap of $^1/_8$in (3mm) and to be free of propolis and wax if they are to work satisfactorily. The vertical alignment should also be checked to ensure that the two springs are central in the case.
b) Two Porter bee escapes per board should be used for rapid clearing and to ensure that if one escape becomes blocked the other one will still be operative.
c) Any clearer board should have an internal bee entrance incorporated in the design with an opening and closing device which can be operated from outside the hive. When wet supers are returned to the hive, the entrance is opened allowing the bees to enter the supers and, conversely, it is closed when they are to be cleared again. It is important that the operating lever allows the roof to be put in place when the supers are off the hive.
d) Approximately 24 to 48 hours are required to clear the supers. The time depends very much on the weather and the flying conditions at the time the board is put on, the better the conditions the shorter the time required to clear.

e) The bee escapes will require cleaning from time to time. Methylated spirit is an ideal solvent for propolis and wax.

Canadian clearer boards: have the advantage of no moving parts to be propolised and go wrong. The salient points of this mode of clearing are as follows:

a) The same time is required (perhaps marginally shorter) to clear; however, if the weather is bad they are not as effective as the Porter bee escapes. The bees seem to learn very quickly that they can return to the supers via the same exit route. The supers must be removed at the latest after 48 hours.

b) An entrance capable of being opened and closed from outside the hive is required identical to the board with Porter bee escapes.

c) If by any chance there may be an odd drone in the supers, they can traverse the exit route without blocking it as would happen with Porter bee escapes.

8 way plastic bee escapes: which is pinned to the underside of the board directly below a suitable hole. There are 8 plastic slots for the bees to reach the brood chamber and there are again no moving parts. The principle is the same as the Canadian clearer board. Our findings are that it works no better than the Canadian board.

Scottish clearer board: has 4 holes in each corner leading to a narrowing channel which reduces to a bee space over a distance of about 3in (76mm). We have found that it works no better than the Canadian clearer board despite what the Scots say about it being superior to any other.

2.16.3 Shake and brush method.

The method appears to be simple and indeed is, if used at the right time.

a) A spare empty super is required, to receive the cleared frames, placed on a roof behind the hive (note the roof is not upturned as most books recommend) with a cover cloth over to prevent any flying bees re-entering the cleared frames. The colony is smoked first at the entrance and then at the top to drive the bees downwards in the supers. One frame at a time is shaken free of bees and those remaining on the frame are brushed off with a feather. The frame free of bees is then placed in the empty super. As one super is cleared so that becomes the receptacle for the next and so on.

b) If the supers are sealed, smoking has little effect on the bees; they are only subdued when they have gorged themselves with honey and when smoked they are not immediately subdued, only driven downwards. After the honey flow has ceased the colony is likely to be more aggressive and will defend their stores. It will be clear that it is not a method to be used by the uninitiated at the wrong time and certainly not in an urban situation.

c) Where should all these bees be shaken? We like to shake them back into the hive rather than at the front, only the bees which we brush off land up at the entrance of the hive. The reason for this is that we keep the super covered with a cover cloth except when a frame is being shaken and if the bees are in the hive, we are in control of the situation and not the bees.

d) The final consideration is when should this method be used? At a time when the bees are not flying to safeguard against robbing being started; this means early morning or late evening.

2.16.4 Other clearing methods.

To provide an overall picture other methods of clearing should be noted:

a) Mechanical blowers usually powered by electric motor which in turn is powered by a portable generator. The super is stood on end and the bees are blown forcibly out of the super towards the hive entrance. Not for the small time beekeeper.
b) Chemical repellents. The three most commonly known are:

Carbolic acid - not used these days.
Butric anhydride - popular in USA.
Benzaldehyde (smells of oil of bitter almonds) - used quite extensively in UK and works well. Should be kept in the dark. The residual crystals which dry out on the clearer cloth are a fire hazard, therefore the cloth should be kept in a sealed tin.

All the above chemical repellents are used by shaking a few drops onto a cloth which is put over the top super and covered with the crown board. When the bees are cleared (a few minutes normally) the super is removed and the cloth and crown board put over the next super below and so on.

2.17 Describe how combs and cappings are dealt with after extraction. Refer to the methods adopted for clearing bees from supers and any treatment of the supers and combs that is routinely carried out before storage.

The requirements for this part of the syllabus are not entirely clear. In the first part it is assumed that the comb referred to is that in super frames. In the second part we are not clear how the methods adopted for clearing supers impacts on these supers before they are stored for the winter. We assume that the Candidate should be able to address the treatment and storage of both brood comb and super comb for this examination.

A noticeable omission in the syllabus appears to be the actual extraction of honey and we advise candidates to be familiar with the various techniques involved.

2.17.1 Describe how super comb and cappings are dealt with after extraction.

2.17.1.1 Cappings.

All uncapping methods require some form of tray or receptacle to receive the cappings and process them. The essential feature of these uncapping trays is to provide a receptacle for the cappings to fall into in order to separate the honey from the wax capping. The two basic methods are cold straining through a suitable gauze or melting the cappings with the honey and allowing them to separate on cooling.

a) The Pratley (or Prattley; there is doubt about the spelling) tray uses the heating principle. It is constructed in stainless steel with a water bath and heating element on the underside. It is expensive and probably one of the worst designed pieces of equipment available. There is no thermostatic control for the heater so one is continually switching on and off. The tray is built on a slope with the thin part of the wedge, and minimum thermal mass, at the place where maximum heat is required. With the result the whole thing clogs up and slows down the extracting process. The separated honey is heated to such an extent it is only good for cooking.

b) The cold straining method seems to be as good as any, the honey is not ruined and the cappings can be washed for mead making or given back to the bees to clean up. Essentially the cappings fall into a perforated tray which allows the honey to drain onto a suitable receptacle below.

c) None of the devices makes provision for locating the frame over the uncapping tray and it is necessary to fix a bar across with a suitable circular hole about $^1/_4$ inch (6mm) deep to rest the lug of the frame in while actually cutting the cappings off.

The wax from honey cappings is of a very high grade for show, for sale in small blocks and for use in cosmetics and ointments and should be processed as follows:

• The cappings will initially be separated in an uncapping tray or similar device with a mesh basket to allow the honey to drain off.

• The cappings can be given back to the bees to clean up or washed to make mead with the washings.

• Finally the cappings are melted and filtered through lint as a final cleaning process. The wax should not be heated above 150°F (65°C) to prevent discolouration. To save the natural colour of the wax, metal containers should be avoided.

2.17.1.2 Super frames and comb.

Super frames with comb require to be cleaned up at the end of each season and it is infinitely easier to do this if they are dry and not sticky with honey. Most beekeepers, therefore, return the wet supers to the hives for the bees to clean and dry them off when they are removed finally for cleaning and storage for the winter.

Each frame should be examined and scraped free of propolis and excess wax before being returned to the actual super which should also be scraped clean, the metal runners wire wooled and greased.

During this examination some frames will inevitably be discarded, the old comb cut out and fitted with new foundation or a starter strip after the frame has been cleaned. The comb from such super frames is processed through a wax extractor and used for making foundation or after thorough cleaning for other uses.

2.17.2 Describe any treatment of brood frames with comb routinely carried out before storage.

Brood boxes and brood frames should be scraped and cleaned as described above for supers and then disinfected with 80% acetic acid. See Section 1.9.3.3.

After fumigation with acetic acid they should be stacked as follows:

a) Mouse excluder with newspaper over.
b) Sprinkle one dessertspoonful of PDB crystals onto newspaper and place brood box over.
c) Cover with newspaper, PDB, another brood box, etc. finishing with a screen and crown board.
d) The PDB should not come in contact with the wax comb where it will contaminate the wax.

2.17.3 Describe any treatment of super frames with comb routinely carried out before storage.

It is unnecessary to fumigate supers with acetic acid and they can be stacked straight away with PDB and newspaper. At the end of the season it is important to get all frames and supers cleaned up and stacked for winter as soon as possible. While the weather is warm, wax moth can do considerable damage.

Note that if supers are stored wet it is necessary to make them bee-proof if they are stored outside, otherwise they must be in a bee-proof shed or room. When the frames are wet, they are not attractive to wax moth so they can be stored without PDB. However, they do become very damp (honey hygroscopic) during winter storage and tend to grow mould during the winter.

2.18 Describe how the colonies are prepared for winter and the timing of carrying out these arrangements.

2.18.1 General.

Preparations for winter should start in August after the main crop has been removed and extracted. There are reasons for this:

a) A colony of bees collects all the stores it needs for winter by the end of July under normal circumstances. If all these sealed stores are removed, sugar syrup has to be fed and this also has to be processed, ripened, stored and sealed; this is difficult for the bees to do on chilly days and nights in autumn, particularly the ripening and evaporating the excess water. It is as well to remember that unsealed stores are likely to ferment and fermenting stores are a cause of dysentery.

b) All colonies require sampling for the adult bee diseases before the colony settles down for winter. If Nosema is present, Fumidil 'B' can be fed with their winter rations. Fumidil 'B' is best fed in the first gallon of winter feed if this is required. If Acarine is present and the crop has been removed, the colony can be treated without fear of tainting any honey. See Section 5.10.

c) Colonies may require to be requeened and it is better to know that the new queen is accepted and laying before clustering starts at 57°F (14°C).

d) All colonies will require to be treated for Varroa mite infestation.

Those colonies that are destined for the heather are prepared before they go with a young queen and hopefully return with a full brood box of stores and a super of surplus honey.

2.18.2 Requirements for successful wintering are as follows:

A sound and weather proof hive.
35lb. of liquid stores.
A young fertile queen.
The colony to be disease free.
Good ventilation while excluding mice.
No disturbance from October to March.

These will be examined in the following paragraphs to understand the importance attached to each.

2.18.3 A sound and weather proof hive.

The principle of this item is to keep the colony dry and must be self evident. However, it is quite surprising the tatty quarters some colonies get landed with; roofs in particular seem to be very often inadequate. Dampness in winter can spell disaster for a colony. Double walled hives have roofs that blow off and a secure method of roping or screwing them down is necessary. If a single walled hive roof has the right clearance between the brood box and the inside dimension of the roof, $^{5}/_{16}$in (8mm) it will not blow off; many do not meet this requirement. Weatherproofing means having the hive off the ground on a suitable hive stand so that the floor board is not permanently damp and can dry out when the weather allows it.

2.18.4 Stores - 35lb (16kg) minimum.

The beekeeper who has to feed his bees before March should not be keeping bees; he has not prepared them adequately for their winter hibernation.

a) After the crop has been removed every frame has to be examined in August and the stores estimated. This is done by eye and by feel; it is surprising how quick and expert one can become at this task. It is essential to know how much a full frame of stores weighs; eg. 5lb (2.25kg) for a BS and 7lb (3.2kg) for a Commercial.

b) Having totted up the total in the colony, a calculation is required to know how much sugar to feed. Honey contains 80% sugar and 20% water approximately. Suppose the colony has 25lb (11.4kg) of stores, then another 10lb (4.5kg) is required to meet the 35lb (16kg) criterion. 10lb (4.5kg) of honey is equivalent to 8lb (3.6kg) of sugar, the amount required to be fed in a suitable solution. It is frightening the number who keep bees and never do this examination and this simple arithmetic. Feed as soon as possible in August.

c) See Section 2.19 for details on feeding and the strength of syrup to feed.

The principle involved in providing the right amount of liquid stores is to provide the fuel that the colony requires to stay alive during the low temperatures experienced during the winter period. This fuel produces heat when metabolised and converted into mechanical energy by the bee. 35lb (16kg) of honey is equivalent to 28lb (12.7kg) of sugar. It is equivalent to the energy produced by a 14 watt light bulb running continuously for 6 months. The arithmetic is as follows for those who may be interested:

100g sugar is equivalent to 1700 kilo joules = 1700 kilo watt seconds

1lb = 454g is equivalent to 1700×4.54 k watt seconds

28lb sugar is equivalent to $1700 \times 4.54 \times 28$ k watt seconds = 216,104,000 watt seconds

For the winter period of ½ year there are $182 \times 24 \times 60 \times 60 = 15,724,800$ seconds

$\therefore$ Size of the bulb = $216,104,000 \div 15,724,800 = 14$ watts.

2.18.5 A young queen.

By observation over the years beekeepers have come to know that colonies winter better with a young queen compared with an older one. This sort of statement will be found throughout the literature but no explanation of why this should be ever seems to be forthcoming. Thus it is impossible to state the principle underlying this particular wintering requirement; the principle has been learnt, by trial and error, over many years by many beekeepers.

It is unlikely to be related to a quantitative problem of queen substance and the threshold amount available to each bee because the colony naturally reduces in size in the winter thereby allowing more queen substance per bee. Perhaps queen substance has other effects on colony well-being which are yet undiscovered. Young queens are likely to lay better than old queens and this could get the colony away to a better start in the late winter/early spring. The other point is what is young; a queen in her 1st, 2nd year etc? Bro. Adam has always maintained that a queen lays better in her 2nd year particularly if she has not been stressed in the first year.

Requeening in the spring with queens bred the previous year and kept in over-wintered nucs will go into winter at approximately 15 months old and carry the colony through winter before being replaced the next March. This system works well. Whether the queen is regarded as old is doubtful, but if she is not replaced at 21 months old, her efficacy thereafter may certainly be expected to taper off quite rapidly.

Our own feeling in the matter, as a result of experience, is to ensure that queens of 24 months old do not lead a colony into winter if this can be avoided. Good queens for breeding purposes can of course be kept in nucs for much longer periods. Nevertheless, the reasons why young queens are better for wintering do seem obscure and the fact will have to be accepted until a more scientific explanation is forthcoming.

2.18.6 The colony to be disease free (in August).

It is vital to sample all colonies for adult bee diseases in August so that treatment may be administered if found to be necessary. Before uniting, which is a common occurrence at the end of the active season, the check for disease is essential.

Examination for adult bee diseases now costs money if the samples are sent to York, and the price per sample discourages beekeepers to use the service. Many counties have organised their own microscopy service and more individual beekeepers are doing their own (this must be good). On the other side of the coin more beekeepers are, for example, feeding Fumidil 'B' to all colonies before winter as a prophylactic against Nosema. Although there are rumblings that this is unlikely to be detrimental in the long run (development of strains resistant to this antibiotic) it would in our opinion be a wrong course of action until a definitive paper has been prepared on the subject by someone with the right scientific ability.

The principle involved here is that diseases are spread in the animal world by close contact one with another making it easy for the disease pathogens to be transferred. Anyone who regularly travels on the crowded London Underground in winter when the common cold is prevalent will readily appreciate the mode of transfer! In winter the bees are in a cluster and at very close quarters to one another for a prolonged period. The importance of the colony going into its clustering mode in a disease free state should now be very apparent.

Varroa treatment takes approximately 6 to 8 weeks (using Bayvarol or Apistan) so it is important that this treatment starts in mid August at the latest so that the strips can be removed by mid October.

2.18.7 Good ventilation while excluding mice.

A colony during winter, if it metabolises 35lb (16kg) of honey, will be required to get rid of approximately 4½ gallons (20.5 litres)of water. This can only be achieved by evaporation. The average rate is 5 pints (2.8 litres)/month or 3oz (84ml)/day. When the colony consumes its sugar stores carbon dioxide (CO_2) and water vapour (H_2O) are produced. The calculation of total water produced which must be dispersed by evaporation is as follows for those readers who may be interested:

1lb honey has c. 18% water	= 3oz (84ml) water
1lb honey has c. 72% sugar	= 13oz (364ml)
To dilute the sugar 50:50 needs other sources	= 13oz (364ml) water, 10oz (280ml) coming from
Metabolising 13oz (364ml) sugar produces 8oz (224ml) water	
∴ Metabolising 1lb (454 g) honey involves 21oz (588ml) water in the form of vapour	
or 1kg honey involves 588 × 2.2ml water = 1.294 litres	

86

20oz (560ml) = 1 pint (568ml)

∴ 35lb (16kg) honey represents 35 × 21oz = 36.74 pints = 4½ gallons water.

or 16kg honey represents 13.6 × 1.294 litres = 20.7 litres = 4.5 gallons

In order to remove this amount of water it will be clear that the basic principle involved is good ventilation throughout the winter. It is more difficult for evaporation to take place in the damp western side of UK compared with the drier eastern side. These are the facts, the best configuration for achieving this evaporation is still being debated in the bee press and still no one seems to agree on the subject.

Our own method, which we have used successfully for many years after experimenting with various approaches, is as follows:

- Heat escaping from the cluster causes the movement of air, warm moist air moves upwards and is replaced by cold dry air at the bottom.
- All entrance blocks have nine $^3/_8$in (10mm) diameter holes drilled in them spaced equidistant apart across the length of the block. This gives a total cross sectional area of c. 1 sq.in (645 sq.mm) to limit the air flow. The block turned through 90° is a normal reduced entrance block with a 4in (100mm) wide slot. The $^3/_8$in (10mm) diameter holes form the mouse guard and are 'kinder' to the pollen collectors in the spring cf the perforated metal mouse guards.
- The crown board is raised about $^1/_8$in (3mm) with matchsticks at each corner. This gives an exit area for air to escape of c. 9 sq.in (5800 sq.mm). The feed hole(s) are covered so that the flow of air is round the outside of the cluster and avoiding the chimney effect directly above the cluster. The roof ventilators now play no part in the ventilation system.
- The mouse guards are put in, usually in September, before the ivy flow and the crown board is raised as late as possible to stop the gap being propolised. It is interesting to note that if there is no ventilation at the top of the colony, in the spring it is usually a mess with condensation and mouldy outside combs. One associated problem is that some of the stored pollen also develops mould and is then useless to the bees. Strains of bee that collect a lot of propolis will get themselves into this situation if crown boards are raised too early.

2.18.8 No disturbance during the winter period.

Once the colony has settled down for winter it should be left undisturbed until the following spring. Experiments have been conducted and it has been found that the cluster temperature is raised quite a considerable amount, up to 10°F (6°C), by say just taking the roof off. Such increases in temperature shortens the life of the winter bees and this manifests itself in spring just when the colony requires all the bees it can muster. Hives should never be sited under trees where the drip of water from the branches above can cause colony disturbance.

The principle involved here is quite simple. By alerting the colony it has to get itself into a state whereby it can defend itself by having bees which are capable of flying and attacking, by stinging, an intruder. For the indirect flight muscles to work they must be at a working temperature, the reasons for this do not concern us here. It will now be clear why the temperature of the colony is raised by disturbance at the cost of increased food consumption and shortening of the winter bee's life.

2.18.9 Other points of interest are:

a) The green woodpecker can spell disaster for a colony if they direct their attention to boring through the side of the hive. They are usually troublesome in very cold weather when they cannot find forage in the hard ground. There are two ways of protection:

- Surrounding the hive with chicken netting.
- Covering the hive with a plastic bag (but note this interferes with the ventilation).

b) It is desirable that a colony has stores of pollen which can then be used when brood rearing starts after the winter solstice. We have never found this to be a problem but there are probably parts of UK where there is a dearth of pollen. The final topping up of pollen stores occurs during the ivy flow in September/October (nb. winter bees are produced by large pollen consumption).

c) Plenty of bees are required for good wintering but making massive colonies by uniting can defeat the object as shown by some experiments done by Dr. Jeffree at Aberdeen University. For a synopsis of his work see Appendix 8 'The size of the colony for winter' in the Green Book.

d) The old adage that bees do not freeze to death but starve to death is very relevant to the wintering problem.

e) The last thing to do is to remove the hive record card from the roof to bring the final year's records up to date and to prepare new cards for the next season.

2.19 Describe methods and reasons for feeding sugar syrup, candy, pollen and pollen substitute.

For simplicity it will be easier to discuss this section under two main headings namely, sugar syrup and candy and then pollen and pollen substitute, the latter being a special feeding requirement when there is a dearth of pollen in the area concerned.

2.19.1 Methods and reasons for feeding sugar or candy.

2.19.1.1 The principles of feeding a colony of bees.

The reasons for feeding a colony sugar are shown below:

a) To provide adequate stores for winter (rapid feeding).
b) To provide emergency stores in the season between colony inspections (rapid feeding).
c) As a means of administering drugs (generally rapid feeding).
d) To stimulate the queen to lay (usually slow feeding).
e) To prevent starvation when the colony is about to succumb (rapid).
f) To enhance wax production and the drawing of foundation and comb (slow or rapid depending on circumstances, eg. a swarm on foundation is fed rapidly).

g) When a colony has an inadequate foraging force, eg. an artificial swarm which is short of stores (rapid feeding) or after spray poisoning losses.

The precautions to take when feeding honeybee colonies:

a) There should be no spilling or dripping of syrup anywhere in the apiary.
b) Precautions should be taken to prevent robbing (reduced entrances and bee-tight hives).
c) Feed should only be administered in the evening just before dark.
d) No sugar syrup should find its way into the supers and be mixed eventually with honey for extraction and sale.
e) Only pure white refined granulated sugar should be used.

Preparing syrup for feeding:

Generally there are two types of mix, a thick syrup for autumn feeding which will be stored more or less immediately and thin syrup for spring or stimulative feeding which is to be consumed without storing. Most of the literature quotes the following:

Thick	- 2lb sugar to 1 pint of water gives 61.5% sugar concentration
Thin	- 1lb sugar to 2 pints of water gives 28.0% sugar concentration
Medium	- 1kg sugar to 1 litre of water gives 50.0% sugar concentration

Since the bee requires a concentration of 50% for it to digest and metabolise the sugar then it is clear which is the best one to use if they are to use it straight away. If sugar syrup is to be mixed with cold water, it will be found difficult to obtain a complete mix with 2lb to 1 pint. The Authors use a mix with cold water of 7lb to 5 pints in an old washing machine (top loader with central agitator). The concentration works out to be 52.8%, less than 61.5% and hence giving the bees a bit more work to do ripening it to 80% for storing and sealing. As we feed for winter immediately after extracting in August, this causes the bees no distress as they have plenty of time to get their larder in the order they require it before the cold nights set in.

2.19.1.2 The most common types of feeders in use.

The requirements of a good feeder are to allow the bees to take the syrup at the rate required by the beekeeper for the management of the colony, while at the same time preventing the bees from drowning in the syrup. Finally, when the feeding is finished, access should be provided for the bees into the feeder so that the bees can clean and dry it up (a job they can do very efficiently given the chance). There is quite an array of feeders available, not all of them meeting the criteria above and many of them being manufactured in materials that can corrode or are difficult to clean. A further disadvantage of some types is that they are capable of being propolised by the bees so that without maintenance they become unusable. The various types commonly available are listed below:

Contact feeders: these come in a variety of shapes and sizes but are all similar in design having a container with a close fitting lid. The lid has a series of small holes or a small piece of gauze through which the bees take the syrup when it is turned upside down over the feed hole or directly onto the frames in the colony. The number of holes regulate the speed that the bees can take the contents. It has the advantage of being cheap and can be readily made at short notice from a bewildering assortment of household containers. The disadvantages are as follows:

a) The bees quickly propolise the small feed holes as soon as it is empty.
b) As the contents are coming to an end, a change in temperature can force the last remaining contents out causing a minor flood of syrup in the hive (usually cleaned up quickly by the bees).
c) They are a bit messy to fill and invert without spilling syrup unless one is very careful.
d) An eke is required in order to house the feeder under the roof.

Round top feeders: are very widely used in UK and are intended to be placed over a feed hole in the crown board. The capacity varies from c.1 pint to 2 or 3 pints depending on the diameter. The height is usually about 3½ inches. The entry is via a tube in the centre and down the outside of the tube to the syrup. The whole of the centre feeding area is enclosed by a removable cover for cleaning. Older versions were made of metal but now most are manufactured in plastic which is better from a corrosion point of view. This type of feeder is easily filled in situ without the bees escaping in the process. Again an eke is necessary.

Miller feeders: were designed by Dr.C.C.Miller in USA and consist of a tray, about 3in (76mm) deep, with dimensions in the horizontal plane exactly matching the external sizes of the brood chamber or supers of the hive it is intended to fit. The entry for the bees is via a slot in the centre extending from one side to the other; again it is provided with a cover to prevent the bees from escaping. The capacity is from 1 to 2 gallons. It allows many bees to feed simultaneously thereby allowing very rapid consumption of the syrup (a strong colony can finish the contents of the feeder in 24 hours.). Construction is generally in wood with all joints glued to prevent leakage. For bottom bee space hives, a bee space is required on the under side of the feeder.

Ashforth feeders: are virtually identical with the Miller feeder except that the feeding slot is placed at one side allowing the hive to be tilted slightly thereby permitting all the syrup to flow towards the feed slot which is impossible with the Miller type and is therefore an improvement. The advantage of allowing all the syrup to be consumed before the tray is opened to the bees for cleaning is that there are no pools of syrup for the bees to drown in.

Bro. Adam feeders: are similar to the Miller and Ashforth except that they have a central entry similar to the Round top type feeders. They are becoming more popular in UK due to some equipment suppliers now manufacturing them. The feeders on the stocks at Buckfast Abbey double up as a crown board (therefore every stock has its own feeder).

All the feeders quoted above are designed for top feeding. Other feeders are available for internal feeding and bottom feeding (which is seldom practised in UK). The internal feeder is in the form of a brood frame with wooden sides and an opening at the top to allow access to the bees. The frame feeder is used for feeding nuclei; the capacity is inadequate for a colony and few would wish to open the colony in order to feed it or refill the feeder. A disadvantage of this type of feeder is the bees propolising the float arrangement (to prevent the bees drowning) in the frame feeder at the bottom when the contents have been consumed.

2.19.1.3 The amounts of food to be fed.

Emergency feeding.

It is necessary to know the amount of food that a colony requires during the season so that, after an inspection, the beekeeper can determine whether it shall require feeding or whether it has sufficient stores to the next inspection. The worst case must always be considered and that is when the colony sends out its foragers and they are unrewarded in their search for food.

A flying bee uses 10mg honey per hour while foraging for an average time of 5 hours per day. If the colony has 13,000 foragers ($^1/_3$ of the total population) and the next inspection is 7 days away, then the colony should have 10lb of liquid stores.
$$\text{ie. } 10lb = (13000 \times 10 \times 10^{-3} \times 5 \times 7) \div 454$$

Therefore, if the colony has less than 10lb of stores it may require emergency feeding if there is no income and the weather is inclement. The amount required is likely to be small, ie. a few pounds.

The same considerations are applicable to nuclei and many a nuc has died out due to starvation because of ignorance of the beekeeper not understanding the little colony's food requirements.

Winter feeding.

We are alarmed and distressed by the large number of beekeepers who either don't know how much food a colony requires for winter or, if they do know, have no idea how to calculate how much it should be fed. The losses each year in the UK due to starvation amount to many thousands of colonies according to a MAFF survey some years ago. We doubt if the situation has changed. If the RSPCA knew more about bees they would be taking some action against the offending beekeepers.

The calculation is a simple bit of arithmetic and the starting point is a colony inspection in August. Each frame in the brood chamber is inspected and the amount of liquid stores estimated on the basis that a BS frame when full and sealed with honey weighs 5lb. A Commercial frame holds 7lb. The calculation is detailed in section 2.18.4 and is repeated here because of its importance.

A strong colony requires c. 35lb to see it through to the spring without feeding early in the new year when stores are used up very quickly. It is often said that a beekeeper who has to feed his colonies in the spring should not be keeping bees! To illustrate the simplicity of the calculation, assume the colony has 25lb of stores after the inspection. The colony requires 35 - 25 = 10lb of additional stores or the honey equivalent thereof. How much sugar must be fed in syrup form to provide the equivalent of 10lb of honey? 1lb of honey contains c. 0.8lb of sugar, therefore, 8lb of sugar should be fed in syrup form. If the colony required 15lb of additional stores then the amount of sugar = 15 × 0.8 = 12lb sugar. It is as simple as that and yet very few beekeepers take the trouble to do the job properly and many colonies starve to death.

2.19.1.4 The types of feed that are fed to colonies of honeybees:

1. The standard feed is white refined household quality sugar either from cane or beet sources (ie. refined sucrose). No brown or unrefined sugar is permissible.
2. It was recommended at one time to feed candy or fondant. It is now used only for special applications (eg. micro mating nucs or the like). If cream of tartar or vinegar is contained in the recipe, both are toxic to bees cf. refined sucrose. It is best not to feed either candy or fondant if it can be avoided.
3. Dry sugar (again refined sucrose) is used by some beekeepers in a tray type crown board usually in the early part of the year supposedly as an insurance policy. It is not recommended because unless water is provided it is extremely difficult for the bees to produce enough saliva to dissolve the crystals.
4. Honey. This should only be fed when it comes from the beekeeper's own apiary and is known to be disease free. Many imported honeys carry AFB spores and are highly dangerous and must under no circumstances be used.
5. Pollen patties are often fed in the early part of the year to provide additional protein where pollen may be in short supply or where colonies are being induced to start brood rearing early. There are two types namely pollen substitutes (fat-free soya flour) and pollen supplements (using trapped pollen; again the source should be from the beekeepers apiaries from disease free colonies). See Section 2.19.2.
6. A comb of sealed honey can often be usefully taken from a disease free colony and used in another requiring urgent liquid stores.

2.19.1.5 The timing of feeding a colony of honeybees.

All feeding of colonies of honeybees should be undertaken only in the evening when it is just getting dark. The reason for this is not explained at all well in most books on bee husbandry. This is curious because it is so important particularly when bees are being kept in gardens at close quarters with neighbours.

The reason is that as soon as food is given to a colony during daylight hours, the scout bees will be alerted and will start roaming the immediate neighbourhood for the source. It seems to be a shortcoming of the communication system of the bees. Presumably a round dance occurs and out go the foragers to seek the source and mayhem starts in the apiary with the attendant possibility of robbing being started also. It seems that the colony has no sure means of indicating to the other foragers in the colony that the source is just above them over the brood chamber in a feeder.

Bees are not equipped for night flying and will not fly in the dark. Hence, all feeding should be done at night. The same goes for putting wet supers back on a colony for drying up after extraction.

2.19.1.6 Other points of interest.

• Each hive should have its own feeder. When feeding starts, particularly in the autumn, all stocks should be fed at the same time.
• There are advantages in combining the feeder as the permanent crown board; it is always

available for use and if it stays on one stock it cannot pass on disease by using it on another colony.
• Open tray feeders with straw or polystyrene chips floating in the syrup are messy and not particularly efficient, the bees often seem to find the 'deep end' and drown in the syrup. Not recommended.
• It is good practice to check the feeders each year for leaks with water before being brought into use.
• Communal feeding has been advocated by some authors by providing a common feeder in the apiary for all colonies to fly to and help themselves. We do not recommend it as the disadvantages far out weigh the advantages. No control can be exercised over what each stock needs and takes. Disease can be spread by this means and you are likely to be feeding someone else's bees!

2.19.2 Methods and reasons for feeding pollen (supplement) and pollen substitute.

2.19.2.1 General.

Pollen is a necessary part of the bee's diet and without it brood rearing would diminish very rapidly and the activation of the bee's hypopharyngeal glands would not occur. Pollen is generally regarded as the source of protein in the bee's diet; however, it not only provides essential proteins but lipids, minerals, and vitamins. Additionally, pollen contains phago-stimulants (Doull, 1974) encouraging the bee to eat it.

Some 35 years ago it was observed that bees collect powdery materials such as coal dust, fine earth, saw dust, rotted wood, etc. in times of pollen dearth (Spencer-Booth, 1960). These observations suggested that it should be possible to feed bees with suitable artificial materials to replace the missing natural diet of pollen. Nb. the collection of tar from the roads when propolis is in short supply.

It was from these beginnings that studies were made, mostly in the USA, of alternatives and how they could be administered to honeybee colonies. Such alternatives are necessary in some parts of the world where there are insufficient pollen sources in the spring (eg. where natural hedgerows have been grubbed out to provide large fields to enable large machines to be used economically) or at other times of the year (eg. honeydew flows where there is no pollen involved).

There are various estimates of the pollen requirements of an average colony in one year, and as early as 1946 Todd and Bishop calculated that 100mg of pollen is required to raise one bee and stated that 20kg (44 pounds) of pollen are required in one year or enough to raise c. 200,000 bees. It is known that the average colony collects far in excess of this amount and conservative estimates are c.57kg (125 pounds). This knowledge led to the idea that excess pollen could be harvested from the hives where plenty was available, the excess pollen being sold as a health food.

2.19.2.2 Definitions of pollen supplements and substitutes.

Pollen supplement is another food not naturally collected by the bees which is fortified with natural pollen.

Pollen substitute is another food which replaces completely the natural pollen collected by the honeybee. Here it should be noted that patties made of pollen substitute were (still are perhaps) called extenders in the USA.

2.19.2.3 Pollen trapping and storage.

2.19.2.3.1 Trapping.

• The earliest trap is attributed to Böttcher in 1941 which was a simple mesh at the front of the hive. Since that time many traps have been designed all with the essential feature of making the bee walk through a grill to knock the pollen loads from its corbiculae. The essential features of a modern pollen trap are as follows:

a) The trapping mesh has dimensions of 5mm made of wire 0.58mm diameter; stainless steel is preferable. Some traps make the bees walk through a double mesh. The separation needs to be 5 to 6mm. As the whole trap forms the entrance to the hive and is generally incorporated under the first brood chamber in lieu of the floor, it is best to have the collecting mesh removable so that it becomes unnecessary to remove the whole trap when trapping is complete.
b) As the pollen loads are knocked off the corbiculae they fall through another mesh into the pollen collecting drawer or container. This mesh needs to be c.4mm or marginally finer allowing the pellets of pollen to drop through while excluding the bees.
c) The pollen drawer needs to have another very fine mesh (fly wire) above its lower surface to collect the pollen thereby allowing a free circulation of air preventing mould forming on the pollen pellets.
d) The drawer should be arranged so that the pollen may be removed from the rear of the hive.
e) Drone escapes (2 to 6) should be incorporated in the design to allow the drones to fly freely.
f) Varroa screens can be incorporated in the design if required to monitor natural mite mortality.
g) It is important that the pollen drawer is made completely waterproof so that the pollen remains as dry as possible.

• Depending on the design of the trap, the amount of pollen entering the hive varies between 10% to 60% of the total amount collected.
• It has been observed that worker bees are often capable of adapting to the trap by collecting smaller pollen loads and thereby increasing the amount entering the hive.
• It is important to ensure that pollen traps are only used on healthy colonies otherwise disease will be spread to other colonies in the pollen used for pollen supplement.
• It is recommended that the pollen is removed (harvested) every 10 days. We would say 2 to 3 days is better to prevent the growth of mould.
• Pollen trapping puts the colony under stress (which is bad news as stress causes disease) and the traps should not be left on permanently. There are various recommendations ranging from 7 days on and 7 days off to 3 weeks on and 1 week off. We prefer the former and to trap when there is a good flow on if possible. Using the 1 week on, 1 week off arrangement necessitates the removable trapping mesh to minimise the amount of work involved.

• The most up to date figures we have seen on the quantities harvested are 5 to 15kg in Arizona quoted by Robson in 1986.
• It was reported from France that if a trap was used on a colony for 40 days during the active season the colony collected 24% less honey and had a 4.4% reduction in brood area (Lavie, 1967); this is no doubt due to there being fewer foragers to collect the crop.
• It is important that a colony fitted with a pollen trap is supervised well; with a good trap, brood rearing is reduced and the colony is put under stress. The rate of increase in pollen collection is proportional to the amount of brood in the colony and also as a result of feeding sugar syrup. It therefore appears to be advantageous to feed syrup while trapping pollen.
• Trapped colonies have a higher percentage of foragers according to Moriya (1966) but this has been disputed by J.B.Free.
• It is stated that care should be taken of pollen which is made toxic by the collection of micro encapsulated pesticides simultaneously with the actual pollen.

2.19.2.3.2 Storage of pollen.

Having collected the pollen for use in pollen supplement it usually has to be stored through to the following year before it is required for use. Poor storage can render it useless as a supplement at a later date. The following point should be observed:

a) The pollen must be as dry as possible. When they are thoroughly air dried, the pellets will lose approximately 20% of their weight and become hard and brittle.
b) It may be stored at room temperature for as long as one year but it gradually loses its palatability and its nutritive value. We have not been able to determine any figures for quantifying these parameters.
c) Pollen is best stored dry and in the freezer at low temperatures to prevent, as much as possible, deterioration in the nutritive value.
d) Storage at low temperatures will ensure that any other form of infestation is killed, eg. pollen mites.
e) Haydak showed the effectiveness of pollen for stimulating the hypopharyngeal glands:

Fresh pollen	100% effective
1 year old	76% effective
2 years old	0% effective.

It was not clear whether his tests were storing pollen at room temperature or at low temperatures. However, it clearly illustrates the importance of not keeping it too long and a very definite shelf life being involved.

2.19.2.4 Preparation of patties.

If patties are to be successful they must support brood rearing, which may be stating the obvious but many substances have been tried and tested before arriving at the most suitable. For example, in Canada fish meal has been tried and is cheap and successful. All are mixed with sugar syrup which not only enhances the making of the patties but also provides sugar which is acting to some extent as a phagostimulant.

Haydak (1967) in the USA has been at the forefront of experimentation and finally found that as a complete substitute the following could be used in the ratio 3 : 1 : 1, fat-free soya flour : brewer's yeast : dry skimmed milk. Adding dried egg yolk improved the nutritional value and the addition of casein (major protein in milk) made it virtually the same as fresh pollen from a nutritive point of view. The whole is mixed into a stiff paste with sugar syrup and made into patties of weight between ½ and 1 pound with a thickness of about ½ inch spread onto plastic film or waxed paper.

Yeast is a protein and it has been stated elsewhere that fat free soya + yeast = fat free soya + skimmed milk; both of which are improved with the dried egg yolk. As a result the standard recipes for patties in the UK has evolved to;

Ingredient	Pollen substitute patties	Pollen supplement patties
Fat free soya flour (%)	75	60
Brewer's yeast (%)	25	20
Natural pollen (%)	0	20
Totals by weight (%)	100	100

Again these are made up with sugar syrup and flattened out into patties ½ inch thick on plastic sheets and stored in plastic bags to stop them drying out.

Pollen substitutes are made commercially and sold under enticing names; they are no better than using the above which will prove to be much less expensive. Since the late 1960s we are not aware of any further work which has been undertaken on artificial materials to replace pollen. Generally, our knowledge is extremely meagre in this direction.

Bees, if given the choice, will select pollen supplement cf. pollen substitute. Both can be made more attractive to the bees by adding small amounts of fragrant scents including honey essence; so perhaps the moral here is to mix the patties with run honey rather than sugar syrup.

We must emphasise again the importance of using fat-free soya flour rather than ordinary soya flour which has a high fat content and is unsuitable for bees. We know of one case where a nucleus succumbed after being fed pollen substitute patties and the only reason we could determine was the doubtful source of the soya flour. Note that expeller process soya bean flour = fat-free soya bean flour.

2.19.2.5 Use of supplements and substitutes.

The patties are messy to make but very easy and simple to use. The patties are put directly onto the brood chamber frames with the plastic film uppermost. It pays to have the plastic much bigger in diameter than the patty to prevent it drying out and becoming crumbly; the bees seem to turn their noses up at the dry bits.

In the UK, patties are usually used in the spring in areas which are predominantly rape growing where there is a dearth of spring flowers to allow the colony to build up naturally. After lifting the crown board, a puff or two of smoke will send the bees down and it easy to place the patties on top and press them down in between the seams of the combs. In February/March a healthy colony may be expected to consume a one pound patty in one week.

In order to build up the colony for the early spring rape flow the patties can be put on the colonies as early as the middle of January; the earlier the better to get the colony as strong as possible before the rape comes into bloom. The major drawback is that swarming problems are likely to occur earlier with such stimulation.

We have no experience using pollen patties at other times of the year, eg. a flow associated with honeydew in July. There would be difficulties feeding the patties with supers on and we would not recommend feeding them at the entrance at that time of the year because of robbing problems.

2.19.2.6 Other points of interest.

It has been suggested that the powdered ingredients can be put outside the hive to let the bees forage normally and return to the hive with loads in their corbiculae. We have not heard of anyone using this method and it appears to be disadvantaged by the lack of sugar as a phagostimulant.

Different pollens vary very considerably in their nutritional value. The honeybee, as far as is known, has no way of determining which is best. Due to the wide range of pollens collected by each colony an average protein value, etc. may be assumed.

2.20 Describe how super combs are stored and the measures taken to combat wax moths.

See also Section 2.17 above which describes how super comb should be handled. Supers may be stored wet or dry. If stored dry, they should be stacked after cleaning and scraping with a sheet of newspaper between each super. On each sheet of paper a desertspoonful of PDB should be sprinkled ensuring that none comes in direct contact with any of the wax comb. The top and bottom of the stack should be made mouse proof with, say, travelling screens or queen excluders.

2.21 Discuss the influence of honey production on apiary procedures.

We had some difficulty understanding the intent of the Examination Board including this section in the syllabus, as all apiary procedures are ultimately aimed at maximising the honey yield whether the apiary is fixed or whether migratory beekeeping is implied. The only management exceptions that we can postulate are when the apiary is used for queen rearing and/or if it is used for producing bees for sale in nucs or as full colonies. We consider that 'apiary procedures' = 'apiary management' throughout the whole season.

On the other hand, we wondered whether the Examination Board are guilty of some sloppy wording which should have been "Discuss the influence of apiary procedures on honey production". Certainly it will be the apiary procedures or apiary management that will have an influence on honey production and not vice versa.

All management procedures are directed at achieving a maximum foraging force (or maximum colony population) for the main flow. This is clearly shown in Appendix 6. The three graphs show one for a healthy colony that doesn't swarm, for a colony which is allowed to swarm and the third graph for a diseased colony. Since the honey yield is directly proportional to the colony population it will be clear in which direction apiary management should be directed.

2.22 Demonstrate how to prepare a colony for moving to another apiary.

2.22.1 The Candidate should note that this part of the syllabus requires the preparation to be demonstrated to the satisfaction of the Examiner. This is likely to be done on one of the stocks selected by the Examiner in the Candidate's apiary. All candidates should be in possession of suitable travelling screens and other necessary items for the examination.

2.22.2 Preparing a stock for moving starts by removing the crown board and replacing it with a travelling screen, preferably with a space of about 1 inch on the underside to allow room for any bees to cluster. When being moved, the entrance should be closed (eg. reduced entrance block with foam pushed into the reduced entrance just before moving) and not restricted with a screen as many books recommend. If light is showing at their normal entrance the bees will attempt to escape at this point and there is the danger of them suffocating in the panic to get out. When being moved, the hive parts have to be secured one with another; this can be done in a variety of ways:

a) Using a hive strap around everything excluding the roof which is always removed for travelling. Two hive straps in opposite directions are safer than one.
b) Screwing plates 4 in × 1in (100mm × 25mm) at an angle of 45º across the joins between floor and boxes and the screen being fastened with screws to the to box. Note that the 2 plates on each side should be angled in opposite directions to prevent movement. This method is considered to be superior to all others but it is more time consuming. It should be used for a major move over long distances, say greater than 50 miles.
c) Spring clips to join the boxes together: these use 3 screws, 2 on one box and 1 on the other.
d) Bro. Adam's method of long bolts through the screen, brood chamber and floorboard.
e) Using hive staples; these are a bit outdated these days and a fine way of disturbing a colony when hammering them home. Definitely not recommended.
f) The entrance block needs securing to the floorboard, the safest way is with two 'L' brackets screwed to the front and to the sides of the floorboard.

Other preparations which are necessary before the actual move are as follows:

a) The site and stands at the new location should be ready to receive the stocks immediately on arrival.
b) Prepare emergency equipment for journey, ie. veil, smoker, fuel, water spray for occasional cooling, spare ropes, wide sticky tape for accidental bee leaks, etc.

c) Make the move during the hours of darkness.
d) When all the colonies are in position at the new site, remove the foam at the entrances to allow free flight. The Publisher (Mr.J.S.Kinross) has advised us that he uses a bathroom loofah in lieu of dense foam! We do not recommend this method for a variety of reasons.

2.23 Describe procedures used for moving a colony a short distance within an apiary and to another site beyond normal flying distance, making reference to the difficulties and dangers involved.

The criteria to be observed when moving colonies of bees from one place to another include optimum distance, vibration, temperature, ventilation and water supply.

2.23.1 The distance.

The distance that bees can be moved is well known, ie. 3 feet maximum or 3 miles minimum, if no bees are to be lost from the colony concerned. Note that it is usually the stock that is moved not the colony (BBKA definition) because it has to be moved in some receptacle or another. The reason for the distance restriction is twofold. Honeybees forage generally up to a distance of 2½ to 3 miles from their hive and have a 'mental picture' of this area or recognise distinctive landmarks within the area and know how to navigate back using these landmarks. Moving their hive within this known area creates a condition whereby the foragers leave the hive in the new position, re-orientate on leaving the hive but while foraging, recognise well known landmarks and return to the old site.

The navigational ability of the honeybee is extremely precise (a matter of a few inches near their own hive). Moving the hive entrance more than 3 feet will create a condition whereby the foragers will not find their hive and will either drift to a nearby hive or cluster at the original position of the hive entrance. The Authors conducted a series of experiments some years ago to test the memories of the bees by moving them to a distant apiary and then returning them to a different site in the original apiary. After two weeks their memories started to fail and all foragers returned to the new hive position in the original apiary. For periods less than 2 weeks, the bees when brought back, continued to return to their original site. Of course during the 2 week period many of the original foragers would have died a natural death and new foragers would have taken their place. The only time that this is not true is when a swarm issues; it can be hived very close to the original site and the foragers do not return to their original hive. It seems that something very curious happens to their memories (?), rather like erasing a computer disc of all its information.

2.23.2 Moving the stocks.

Moving the stocks involves observing some simple rules:

a) Place foam in reduced entrance and then remove roof.
b) Place the stocks with the frames in a fore and aft direction so that frames cannot swing if emergency braking or stopping is required en route.

c) Ensure all stocks are roped down securely before starting. Stop after 15 minutes and check all is secure (tension up if required).

d) Corner at slow speed to minimise frames swinging.

e) The stocks should be moved preferably during the hours of darkness arriving at the destination about daybreak. If they are moved during the day over heating must be watched carefully and cooling applied (say every hour) with water spray if necessary.

f) If they are being moved on a trailer ensure that it has a spare wheel.

g) On arrival, set up all the stocks in their final positions, replace all roofs and immediately remove foam at reduced entrances.

h) Next day remove screens and replace crown boards.

2.23.3 Vibration.

Vibration excites bees and if they are closed up in transport the temperature increase would be dangerous if insufficient ventilation and cooling were not provided. During transportation by vehicle there will be a continuous vibration keeping the colony in a state of agitation and high temperature. It will therefore be clear that vibration in general is closely allied to temperature and ventilation. In order to minimise these adverse effects, stocks should be handled with care during the loading and off loading process.

2.23.4 Temperature and ventilation.

Temperature and ventilation go hand in hand and, of course, are allied to vibration. Because of the rise in temperature when a colony is disturbed it is necessary to provide adequate ventilation when moving bees. If very strong colonies are to be moved then it may be advantageous to provide additional space by adding another super as well as providing the ventilation screen. Even these precautions when moving strong stocks during the day in warm weather, may be insufficient to prevent dangerous temperature rises, enough to melt wax comb and drown the bees in honey. Spraying the colony with water through the ventilation screen will be required as part of the operation.

2.23.5 Water supply.

Water may be required en route as indicated above and it will be obvious that a regular water supply will be required by the colony when it arrives at its new location.

2.23.6 Other points related to moving bees.

a) Bees should only normally be moved during the flying season, the winter cluster should not be disturbed.

b) Continual movement of bees, for say pollination purposes, puts them under stress and stress is the forerunner to Nosema.

c) It is better to move a stock of bees some days after it has been inspected in order to allow time for the bees to re-propolise all the seals which had been broken. This minimises internal movement of frames etc.

d) Travelling screens should be constructed of a mesh of 7 to 1 inch of a wire gauge c. 28 SWG (standard wire gauge).

e) Colonies being moved to a new site should have a 10 day supply of stores in the brood chamber.

Stories of moving bees are legion, most of them involve the bees escaping and someone being stung. We find no amusement in these stories and believe that if bees are moved properly and the common sense precautions outlined above are taken, then no bees will be lost and no one will be stung. We consider that any bees escaping en route is due to negligence on the part of the beekeeper.

2.24 Describe the procedures used to prepare a nucleus colony. Discuss the many uses for a nucleus colony.

2.24.1 Definition of a nucleus.

Before looking at the methods of making a nucleus (popularly referred to as 'nuc' in the singular and 'nucs' for the plural of nuclei) it would be as well to examine the definition of a nuc to understand what has to be made. The BBKA standard is that it shall be "a colony occupying not less than three BS (British Standard) combs, 14 in × 8½in (356mm × 216mm), of bees and not greater than five BS combs with the brood (eggs and worker brood) area not less than half the total comb area". As this is a standard for sale it also covers the amount of food and that all the frames should be well covered with bees, etc., etc. The standard for a colony is six BS frames and greater. There seems to be no formally accepted definition of a nuc in past literature but the BBKA standard above will serve our purpose reasonably well as a target to aim at when making a nuc. The number of bees on a well covered BS frame ranges from 1000 to 2000 bees; say an average of 1500 bees (750 on each side). With 3000 bees it is unlikely that any part of the comb will be seen. It follows that using these figures the number of bees in the minimum sized nuc (3 BS frames) should be about 4500. For the maximum sized nuc (5 BS frames) 7500 bees would be required, c.1.5lb (0.7kg) in weight. Nucs on other sized frames would be proportionally sized but with very large sized frames the minimum could not be reasonably less than 2 frames to allow a brood nest temperature to be maintained between the two combs. Other nucs, such as 'mini nucs' and 'micro nucs' are very specialised for queen mating and are beyond the scope of this book. However, their existence should be noted and that they would not fit the definition postulated above.

2.24.2 The nuc box.

This is in effect a miniature hive but with some specialised requirements. To consider the principles involved, the definition of a nuc as above will be used. The requirements are as follows:

• It shall be capable of holding 5 BS frames plus a dummy board.
• The inside width to accommodate this shall be 5 × 1½in (38mm) = 7½in (191mm) plus ½in (13mm) for the dummy board plus ¼in (6mm) clearance making a total of 8¼in (210mm). If Hoffman spacing is used, 1³/₈in (35mm), then the clearance will become ⁷/₈in (22mm). If 1½in (38mm) spacing is used it would be preferable to increase the clearance; it is soon used up when

inserting a Butler cage for queen introduction.

• A dummy board is essential in a nuc because it will only contain 3 frames when it is initially made.

• The entrance arrangement is important; it should be capable of providing plenty of ventilation but be capable of being restricted to prevent robbing. In-built mouse guards are useful if the nucs are used for over wintering new queens for the spring.

• The crown board requires two large ventilators at the back and front covered with wire mesh (8 per inch approx.). A feed hole is necessary to accommodate the nuc's feeder.

• Every nuc box should have its own feeder permanently placed on the crown board. Our preference is for mini Ashforth feeders holding about 1 pint of syrup; they cannot be bought and we have to make our own. If a frame feeder is preferred then the inside dimensions of the box will have to be greater than above; these are not recommended because of opening up the nuc to feed which is bad practice when the nuc is being used for queen mating.

• The roof has to be dimensioned to accommodate the feeder; if the feeder is made too large the roof becomes disproportionate in size and creates too much windage. The most essential features of the roof are that it should be absolutely bee tight (because of feeding) and the ventilators should have an area equal to that of the crown board ventilators. Provision should be made for keeping the record card in the roof so that it is readily available.

• The depth of the box requires a clearance of 1 inch (25mm) below the bottom of the frames to accommodate a queen cell protruding from the bottom of a frame. It does not happen very often but it is useful to have.

• The easy mobility of a nuc is essential and each nuc should have its own travelling screen easily and securely fastened when required.

• Provision should be made to raise the nuc off the ground with two parallel bars on the underside dimensioned so that the nuc has a forward slope, about ½in (13mm), to drain off any condensation when overwintering and to have an air flow under the nuc to keep it dry.

It is impossible to purchase a nucleus box to the above specification and it will be a case of making your own as a DIY activity or modifying one made by the bee supply manufacturers. We put up for many years with commercially made equipment which in many cases would not perform efficiently the task it is intended to do. It is a great pity that others new to the craft have to go through the same loop because designs are not improved.

2.24.3 Making the nuc.

The essential components for making a nuc are a queen, bees, food (honey and pollen) and emerging brood. If the nuc is to be used for mating then a QC can be given to the nuc in lieu of a queen. A nuc can be made from:

     a) a single colony,
     b) from two colonies or
     c) several colonies.

There will be a difference depending whether the nuc is to remain in the same apiary or whether it is to be moved more than 3 miles away, the latter being a much easier task.

Method 1. Three frame nuc (to be transported away). From the parent colony find the queen and cage her. Select two frames of emerging and advanced brood with attendant bees and place in the nuc box. Then select a good frame of food containing fresh pollen and liquid stores, again with bees and also put into the nuc box. Find now a really well covered frame of bees and shake the lot into the nuc. The old queen can be released into the nuc or a new laying queen can be introduced in a Butler cage. The dummy board should be inserted and the space filled on its vacant side with a piece of foam for travelling. Fix the travelling screen and move to the new site immediately. On arrival at the new site open the entrance (reduced) and let the bees fly. If the nuc was destined to receive a QC then the nuc would be transported queenless and without the QC which would be put in at the new site. The nuc should be fed straightaway. Bees will be dying every day through natural causes and these will be replaced by the emerging brood. If a new laying queen is introduced it will be 21 days before any of her progeny hatch out and longer if the QC has to hatch and the virgin to mate before laying commences. During this time the little colony is unbalanced and in a delicate state until it becomes established; therefore, it must be treated with great care to prevent it being robbed. Continual feeding may be necessary.

Method 2. Three frame nuc (to remain in the same apiary). Proceed as in method 1 but ensure that the frame of liquid stores is virtually full so that the made up nuc can survive without feeding for about 4/5 days. The additional bees shaken into the nuc will be greater in this method to allow for any flying bees returning back to the parent colony. Before shaking into the nuc, lightly shake the frame in the parent colony to get rid of the older bees and then shake the rest into the nuc. Do this with three frames and then introduce the queen and place in a new position in the apiary, out of the flight path of other hives, with a reduced entrance lightly closed with grass. Check after 4/5 days and then feed as required.

If nucs are made up as above taking frames and bees from different colonies, it is prudent to spray each frame lightly with a very weak water and sugar syrup leaving the frames well apart in the nuc box and exposed to the light. The bees that are shaken in should also be lightly sprayed. Finally, slowly bring the frames together after smoking well; it is unlikely that any fighting will occur following this treatment. Alternatively, all the frames less bees can be taken from one or different colonies and the bees from another colony. If it is possible to avoid mixing bees from different colonies, then this should be done. Needless to say, nucs should only be made from disease free colonies.

2.24.4 An account of the various uses of nuclei.

Nuclei are an essential part of modern apiary management and are probably more useful for teaching purposes than a large full sized colony. Manipulating a large colony can be a daunting experience for the newcomers to beekeeping and their initiation should always be on a nuc. The small colonies should form part of every beekeeping establishment whether it be a commercial honey producing organisation or a small amateur beekeeper with a couple of hives. The number of uses to which nuclei can be put is really quite remarkable, the important ones are listed below:

  1. Queen mating (these nucs can be quite small, nb. mini nucs).

2. Establishing and building into a full colony.
3. Increasing stocks and replacing colonies.
4. Swarm control.
5. Keeping spare queens and breeder queens.
6. Assessing the queen's offspring.
7. Drawing worker comb.
8. Observation hives.
9. Requeening large stocks.

Queen mating. Probably tops the list of uses and is probably the most complicated. The size can range considerably from the micro nuc with only a few dozen bees to a 5 frame nuc on BS frames. The important feature is that if no brood is present the little colony is prone to abscond. The presence of brood creates conditions which are favourable to the acceptance of a queen cell and there will certainly be no absconding as a mating swarm. Bees will not leave brood which needs tending whether it be young or emerging. Introducing a queen cell to an established nuc always seems to cause confusion with many beekeepers. How long should the nuc be left queenless? There are the following possibilities:

a) Remove the queen and introduce a ripe QC (14 days old or greater) straightaway. There is a fair possibility that the cell may be destroyed; a cell protector (or a bit of sellotape) is a good insurance policy.
b) Leaving the nuc queenless for about 2 hours gives a high acceptance success rate. Some advocate feeding at the time the cell is introduced but the rationale for doing this is obscure.
c) Leave queenless for 7 days and then destroy emergency QCs before introducing the ripe QC. In our experience this is 100% successful.

Making up a nuc specially for mating purposes is the final option. Care must be taken to ensure the brood is only advanced or emerging. If it is left for two days in a queenless condition the ripe QC will be accepted without trouble. Usually 100% successful.

Establishing and building into a full sized colony. This is the ideal way for a beginner to start beekeeping. The ideal time is to take possession of the nuc in March/April and, of course, this will be an overwintered one. It will be capable of being built into a full sized colony and a surplus obtained during the first season. If the nuc is obtained in June with a current year queen, the build up is unlikely to lead to a surplus during the first season.

Increasing stocks and replacing colonies. Even with the best management and bee husbandry, occasionally stocks are lost during the winter for a variety of reasons. Overwintered nucs are ideal for replacing such losses and will provide a crop during the year of replacement. If stocks are to be increased, then new nucs will have to be made up during the season. The normal time for this is May/June when queens can be also reared and the colonies are strong enough to provide the bees for making the nucs.

Swarm control. Removing bees and brood from strong colonies to make nucs is an effective method of swarm prevention by reducing the colony population. If QCs are present in the parent colony, one of these may be usefully used in the nuc when it is made up. The danger of perpetuating a swarming strain must be taken into consideration with this particular use.

Keeping spare queens and breeder queens. All beekeepers should have a spare queen available for emergency purposes. This means maintaining an overwintered nuc or two in case one is required early in the year when it would be impossible for a virgin to mate due to drones not being available. The life of a breeder queen can be extended by keeping her in a nuc and thereby severely limiting the extent of her egg laying. In fact the genetic material (eggs and larvae) for queen rearing can be obtained directly from the breeder queen in the nuc. Breeder queens can be kept for up to 5 years in this way.

Assessing the queen's offspring. Because of the very widespread problem of bad temper, we consider it essential that new queens are assessed in nucs prior to being introduced into colonies. It is easy to deal with bad tempered bees if they are in small numbers. Other characteristics are observed such as laying pattern, nervousness, amount of propolis collected, etc.

Drawing worker comb. Small colonies produce little, if any, drone comb when compared with large colonies. A nuc will always draw worker comb irrespective of whether foundation has been provided. Therefore, old comb with the drone comb cut out can be given to nucs for repair as well as giving them foundation. All our 5 frame nucs are given 1 or 2 frames of foundation to pull out every year.

Observation hives. These contain only two or three frames of bees and are therefore stocked from a nucleus. A greater use could be made of observation hives for learning and teaching than is done at present. The observation hive can be stocked from a nuc and given starters instead of foundation to observe comb building in progress.

Requeening large stocks. If a queen is purchased or obtained from another source and has been out of the hive for some time, it is best that she is introduced initially to a nuc (there is a better chance of acceptance in a small colony). When her laying has normalised in the nuc, then she can be introduced into the large colony. For successful queen introduction, it seems that the old and the new queen must be in the same physiological state. An alternative method is to make a nuc from the colony to be requeened, introduce the queen to the nuc and when laying normally the nuc is united with the parent colony, after first removing the old queen, thus bringing it back to its full strength. We recommend, as an insurance policy particularly if the queen is yellow and the colony is black, that the queen is recaged for a day when the nuc is returned to the parent colony.

2.24.5 Other points.

• Chalk brood always seems to be a problem with nucs until they become established as well balanced colonies, albeit small ones. The trigger is of course temperature, protein and CO_2 stress. When the nucs are established and good ventilation is provided as per the spec in Section 2.24.2, it has been our experience that they seem able to keep it at bay.
• Nucs made up as 3 frames in early June with QCs, build up to 5 frames and generally collect

enough stores to feed themselves for winter. They overwinter well with the young queens and provide the replacement queens for the spring.
• Samples for adult bee diseases should be taken from the nucs and treated in a similar manner to full colonies.
• Nucs should never be allowed to raise a queen from their own emergency QCs; scrub queens will result.

2.25 Describe the actions required to deal with a vicious stock of bees.

2.25.1 Prevention is always better than cure. It is far better to avoid rearing bad tempered bees rather than having to deal with them when they are troublesome. Actions to achieve this aim are, by and large, common sense and are listed below as follows:

a) Assess all new queens in nucs prior to introducing them into honey producing units. See Sections 2.12 and 2.24.
b) Move the stock to a 'bolt hole' at the first sign of bad temper. It is essential for all beekeepers to have an isolated site (= bolt hole) where rogue stocks can be taken at short notice.
c) In lieu of b) above be prepared to kill the stock at short notice.
d) Always have a spare queen available during the active season from March to October.
e) Under normal circumstances handle your bees without gloves; if you can do this your problems of bad tempered bees will be a thing of the past.

2.25.2 Requeening bad tempered colonies.

Always the problem is to find the queen and it will be a queen that is unmarked and the colony owned by an indifferent beekeeper. They always squeak when in trouble and the situation is desperate. Remove the colony to a bolt hole, if at all possible, is the best advice.

It is always best to have a look at the colony quietly to see if it is really bad tempered. We have had many calls only to find that we could handle the colony without gloves and most of the trouble is the way the beekeeper manipulates the bees, using bad smoker technique, clumsy handling, etc.

By moving the colony. Early morning move the colony away from its normal position (about 8 feet) and put an empty brood box with about six combs in it to catch the foraging bees on return. Remember about 33% of the colony are foragers and sufficient comb should be provided. At mid-day when all the foragers have gone find the queen and introduce the new one. Next morning swap the boxes over returning the hive to its original stand and the foragers in the place where the hive was temporarily placed. Remove the empty box of frames the evening of the second day. Four days later release the queen if the release is to be supervised or remove the cage if released by the bees.

By using chloroform. This is only for the really bad stocks! The colony that sends the bees downwards into your 'wellies' and finds every crack in your armour. Kit up well and carefully because the returning foragers are the problem but these can be avoided by shifting the stock as above.

The method is due to the late Harrison Ashforth, CBI for Cornwall. It works well and we have used it on quite a few occasions on other people's bees. About 1 fluid ounce of chloroform is required, two pieces of corrugated cardboard about 3" square and a piece of foam to seal the entrance. The method is as follows:

• The entrance block should be in the colony. Drive the bees back from the entrance with smoke.
• Pour one teaspoonful of chloroform onto one of the pieces of cardboard, push into the entrance and seal with foam or a rag.
• Do exactly the same with the second piece of cardboard putting it under the crown board; the feed hole should be covered to keep the fumes inside.
• Wait for 2 to 3 minutes.
• You then have about 10 minutes maximum to work and find the queen.
• Remove her and run another one in directly from a cage lying on the top bars.
• As soon as the new queen has walked in close up and open the entrance. The job is complete.

The chloroform affects the central nervous system of the bees and disorientates them. We assume the mechanism is that in a disorientated state they cannot recognise their own queen and immediately accept the new one. The method has, so far, never failed and Harrison Ashforth also reported similar results with no failures.

We have heard that chloroform is difficult to obtain. If you tell your local chemist what you want it for and you are prepared to settle for a small quantity, we understand that you may well be lucky in your quest.

2.26 Demonstrate the procedures used for uniting two colonies and the precautions that need to be taken.

2.26.1 General considerations.

There are various points concerning bee behaviour and beekeeping practice which are of interest before considering the possible methods of actually uniting, these are:

a) Both colonies must be disease free; the spread of disease is caused more often by the beekeeper rather than by any other mechanism.
b) In beekeeping literature mention will be made of 'colony odour' and 'hive odour'. Butler (of queen substance fame) postulated that colony odour is genetically produced and each colony has its own characteristics. On the other hand, Bro. Adam was of the opinion that there is no such thing as colony odour but that there is a hive odour which depends entirely on the materials of the hive and the income (nectar and pollen) which in turn depends on the weather. The hive odour is carried by the individual bees. No one has disputed the concept of hive odour but it has not been subjected to any scientific experiments or proof.
c) During times of dearth there are many guard bees at the entrance of a colony, some of these being potential foragers if forage was available.
d) When there is plenty of forage and a flow on, there will be virtually no guards at the entrance. It is likely that all the colonies in the same apiary are working the same crop and the hive odours

are likely to be very similar. Under these conditions, drifting bees are accepted in another colony without challenge or fighting.

From the considerations above it is clear that the best time to unite colonies is during a flow. Feeding, particularly with a scented syrup, when there is a dearth is the alternative solution although it is not too easy to feed both colonies separately during the uniting process and feeding both separately beforehand is usually the order of the day.

2.26.2 Methods of uniting.

There are a variety of ways of uniting; the more important methods and variations will be described. These are:

      1. Newspaper method.
      2. Direct uniting.

Newspaper method. This method is probably the most widely used and is generally very reliable and successful in use. The principle involved is very simple; a queenright (QR) colony and a queenless (QL) colony are joined together with a sheet of newspaper between them and the bees chew the paper away and intermingle slowly and hence unite. The paper is deposited outside the hive in the course of the next 24 hours.

• The two colonies to be united have to be brought adjacent to one another (see Section 2.23) with their entrances in the same direction.
• The manipulation of uniting colonies should be done in the evening when both colonies have virtually finished flying. The reason is obvious, if the bees are flying then some of the returning foragers will be returning to the entrance of a foreign colony and fighting is likely. Once fighting starts more guards are alerted and then all bees from the other colony trying to enter will be involved. This simple precaution is seldom, if ever, recommended in the literature on practical beekeeping.
• The newspaper requires 3 or 4 pin holes made in it to help start the process of paper destruction. This can be done with the corner of the hive tool blade if care is used. It is useful to cover the paper with the queen excluder to stop it blowing around during the manipulation. Note the requirement to remove the queen excluder the following day after uniting to release any drones above it.
• Prior feeding is required if there is no flow on.
• One colony must be dequeened, the first part of the manipulation. Some books have suggested in the past that the two queens will fight it out and the younger queen will succeed. There is no definite proof that this is so and the possibility exists that the surviving queen may be damaged in the fight. Our advice is do your own selection and be sure of the result.
• Now comes the last, but vexed, question of which goes on top and which goes below? There is the queenright (QR) colony and the queenless (QL) colony and either may be the strong (STR) one or the weak (WK) one. Consultation of 4 books which are recommended reading for the BBKA exams gave the following result:

	BK1	**BK2**	**BK3**	**BK4**
QR or QL on top	QR	QL	QR	---
WK or STR on top	WK	---	STR	WK

The curious thing is that although the authors were recommending a particular approach, not one of them explained why their way was presumably the right way and whether either of the two conditions take preference. If anything preference would be given to the strong colony being above the weak colony on the basis that the weaker would have the minimum guards at the entrance to oppose returning foragers. A case could be made for having the QR colony below with a queen excluder over on the basis that the arrangement can be left for 3 weeks to allow all the brood to hatch in the upper box; the top brood box can then be removed. We are of the opinion that it does not matter which way round they go and any combination will be successful if 3 criteria are observed, namely:

• Dequeen one colony.
• Do the manipulation in a flow or with colonies fed for 2 days before the manipulation.
• Do the manipulation just as it is getting dark.

It is not clear where or when the newspaper method originated but it is simple and effective. In some ways it is similar to using a screen between the two colonies for a few days before removing the screen to allow the bees to unite. This method has the disadvantage of requiring a separate entrance for the top colony which is closed when the screen is removed.

Direct uniting. This is usually undertaken with small colonies (eg. 2 nucs to be united) which may, in total when combined, fill a single brood box. The colonies are brought together, one dequeened and again the operation undertaken in the evening as follows:

• Each frame with bees is removed one at a time from the colonies, dusted with flour or sprayed with a weak syrup and placed in the new brood chamber.
• The frames are taken alternately one from one colony and then one from the other and placed in the new brood box also alternately so that there is a complete mixing.
• Care should be taken not to split the brood nests which should combine to make one large one.
• Finally, the bees are heavily smoked and bumped around to create confusion and the colony closed up.

The flour or the syrup gives the bees an immediate job to do and fighting is non-existent or else minimised.

The words of Bro. Adam on the subject of uniting are of considerable interest; exposure to light has a calming effect on bees and when they have been exposed for some minutes, they will peaceably unite without any other precaution throughout the whole season. We follow his wisdom with small colonies and nuclei but use the newspaper method for larger colonies. A further variation on the direct uniting method is to place the frame with the queen and bees in the new brood box and then shake all the other bees from both colonies in front of the hive, placing the empty frames in the brood box. The shaken bees are sprayed with syrup or dusted with flour and allowed to return to the hive. This method is not one that is recommended these days, the job can be done with less confusion and uproar in the apiary.

2.26.3 Other points relating to uniting.

• Swarms can be thrown together (queens and all) into the same hive within a few days of one another without fighting. When there is a surplus of swarms it is a good way of dealing with them. We have bumped up to 3 swarms all together on the same day and had two brood boxes of foundation drawn out in 2 weeks; when the flow starts such a unit will collect a surplus.
• Some books state that uniting when there are no drones about is a bad time for this operation; the rationale is not understood. Uniting colonies before winter is a classic time for rationalising the apiary.
• It is a well known observation that a strong colony will collect more surplus than two weak ones; it is important to ascertain the reason for weakness. If it is disease or poor queens, then uniting will not alleviate the problem. In general the honey produced by a colony is directly proportional to the size of the colony when the colony size is over about 25k bees. The following figures were derived in New Zealand and demonstrate well the benefits to be obtained by having large colonies.

SIZE OF COLONY(bees)	HONEY PRODUCTION (kg)
10,000	4
20,000	14
30,000	23
40,000	32
50,000	41
60,000	50

Note the difference between 10k and 20k and the linearity above 30k. Therefore, there are advantages to be gained by uniting colonies before the main flow and disadvantages in the autumn by uniting and ending up with a colony too large for the optimum wintering conditions.
• It has been suggested that a colony of laying workers can be united to a QR colony. Such colonies are virtually impossible to requeen *. We disagree that uniting is a solution because, if there are laying workers, the colony will have been QL for 3 weeks or more and all the bees will be old ones. If united satisfactorily they will die off quickly by natural causes and the recipient colony will derive little gain from the addition.

> * Work in France in c.1988 indicated that it is possible to requeen colonies of laying workers by dipping the queen in a solution of royal jelly (70%) and water (30%) and introducing them directly. The success rate claimed is greater than 70%. We believe this to be only of academic interest and has no application from a practical beekeeping aspect.

2.27 Demonstrate how beeswax is recovered with reference to the actual equipment used.

It is to be noted that the Candidate is required to demonstrate to the Examiner how beeswax is recovered and to have prepared samples of clean moulded wax for sale. These samples of wax should not weigh less than 25g (one ounce blocks).

During an average season a beekeeper will expect to collect about one pound of wax from each stock of bees during the year. This wax will include the cappings from the sealed combs of honey, the brace and burr comb removed from the frames during inspections, any wild comb built by the bees and old brood combs which are discarded. The cappings are of high quality and need very little treatment to clean them for use. They should be used solely for show wax, making cosmetics, candles or high quality wax blocks for sale. Quite a considerable amount of brace and burr comb is collected during colony manipulations and inspections. This wax should be lumped in with the old combs for rendering. These last mentioned sources contain wax contaminated with propolis (which cannot be removed); it is only suitable for re-use as foundation, candles, etc. after rendering and cleaning. There is no way of cleaning wax at home that is comparable with large scale commercial operations with heated pumps and filters.

2.27.1 Small scale methods of recovering beeswax from cappings.

• The cappings will be initially separated in a decapping tray or similar device with a mesh basket to allow the honey to drain off or spun again in the extractor using a fine mesh basket or nylon bag.
• The cappings can be given back to the bees to clean up in an Ashforth type feeder (only do this in the evening after the bees have finished flying), washed to make mead with the washings or washed with soft water and drained, dried and set aside for making show wax.
• For show wax remove any discoloured pieces of wax.
• The cappings, free from honey, are melted and filtered first through lint and then filter paper as a final cleaning process. The wax should not be heated above 194°F (90°C) to prevent discolouration. To save the natural colour of the wax iron containers should not be used. The containers for moulding should be kept solely for this purpose.

2.27.2 Processing old comb.

When the wax in the old combs melts it adheres to and saturates the old larval skins and except by pressing or centrifuging at high temperatures some wax will inevitably be lost. These two methods are generally unsuitable for home operation. There are two suitable methods for home use, these are the solar wax extractor and the steam boiler. Old combs contain very little wax. If only a few combs are involved or there are disease pathogens present then the combs are best destroyed by burning.

The steam wax extractor.
This is a stainless steel boiler with a mesh cage suspended inside with a drain at the bottom for the wax to run off and the steam to escape. The device has a water reservoir which is converted to steam which melts the comb and wax inside. The boiler has an air-tight lid. It can be driven by gas or electricity. The largest steam wax extractor seen by the authors is at Buckfast Abbey. Here the wax steamer takes an entire brood box and frames.

The solar wax extractor

This is probably the best device for recovering wax. A second melting of the old comb wax is usually necessary. Heat the broken wax in a saucepan of soft water until all the wax has melted. Leave the wax to cool and solidify on top of the water (Specific Gravity of beeswax c 0.95). Any dross, 'slumgum', can be scraped off the bottom of the cold block of wax. Further cleaning by filtering may be necessary if cleaner wax is required.

The M &G extractor.

This piece of equipment is still offered for sale by the commercial appliance suppliers at about £110. It works well with care and controlled heating. It consists of a metal drum, old comb is put inside and filled up with water. A filter (fine nylon) is tied across the open top. The water and wax are heated. Around the outside and as part of the device there is a large rim to catch the contents as they are pushed through the filter. The rim has a spouted outlet for the wax. The wax is forced out through the filter onto the rim by hydraulic pressure when water is poured into the high spout, the bottom end of which is connected into the lower part of the drum. If care is not used it is claimed that the ceiling is likely to receive a wax treatment. This method is definitely an outdoor activity.

Using an old 44 gallon oil drum.

The old beekeeping books show this outdoor method. The tank is filled with water and heated by wood/gas. All the old combs are placed in a sack which is kept below the surface of the boiling water with a strong stick. As the wax melts it passes out of the sack and rises to the surface of the water. Rainwater is usually recommended! When all the wax has passed out of the sack the wax is allowed to cool and solidify on top of the water.

2.27.3 Other points.

• Most appliance suppliers will purchase rendered beeswax or exchange it for equipment or foundation.
• When handling small amounts of wax especially on domestic appliances it should be remembered that wax is very inflammable. Use double saucepans made of aluminium or stainless steel. Should any wax spill wait until the wax has solidified before attempting to remove it. Covering the top surface of the oven with aluminium foil is a wise precaution to take before undertaking any wax melting.

** ** ** **

NATURAL HISTORY AND BEHAVIOUR

The Candidate will be able to describe the following and explain their relevance to practical beekeeping:

3.1 The different races of honeybees and their characteristics.

3.1.1 How the variable temperaments of bees arise.

In order to address this part of the syllabus it is necessary to have a clear understanding of why the temperaments of our honeybees are variable and how this arose. There is a very wide range of temperaments from the very docile to the very aggressive or, put another way, from those that exhibit little defensive instinct to those with a very strong defensive instinct whether the colony is disturbed or remains undisturbed.

By various scientific techniques, including palaeontology, it is generally agreed that the origin of the honeybee was in that part of Africa that is now known as Kenya somewhere between 20 and 30 million years ago. This was before India split off from Africa and before the Red Sea and Rift Valley were formed. The land mass now called India carried with it some of the earlier bee-like insects which developed into the Eastern species of bees, ie, *Dorsata*, *Cerana* and *Florea*.

Reference now to the diagram of the 'Migration of the Honeybee' in Appendix 7 shows the origin in Kenya and the main migration routes marked with double line arrows northwards, southwards and westwards. Major races developed notably *A.M.capensis* in the Cape of Good Hope area, *Adansonii* in the equatorial strip, *Fasciata* (the Egyptian bee) in the north east and *Intermissa* (the Tellian or Arab bee) in the north west.

The Tellian bee is considered to be a major race from which many other strains have developed. In its native part of northwest Africa it developed in a very hostile environment and presumably adapted to these conditions by having to defend its colonies against determined predators (eg hornets, etc.) It migrated northwards before the last melting of the icecap c. 10,000 years ago and established the well known races in NW Europe such as the Iberian bee, the French Black bee and the English bee. It migrated as far north as Finland in latitude 60° N.

Similarly, the other major race (*Fasciata*) migrated northwards also providing the Italian bees (*Ligustica*), Greek, Caucasian, Carniolan, etc. After the ice cap melting the UK was cut off from the continent and the Mediterranean was formed isolating Europe from Africa. At that time we had a discrete race of bee in the UK now known as the English Black bee or the English Brown bee. It was decimated by disease in the early 1900s (nb. Isle of Wight Disease or Acarine) and the Government of the day offered subsidies to beekeepers to import bees from the Continent. The result was that most of the races in Europe were imported and over the last 50 to 80 years these races have interbred and we now have a hotchpotch of mongrels with a very mixed pool of genes.

All the bees with their origins or stem emanating from the Tellian bee have to a greater or lesser degree the defensive instinct while other bees from Italy, Greece, etc. have a very weak defensive trait. Breeding can only be true with pure strains while breeding with mongrels is known to be an erratic and unpredictable procedure.

Herein then lies the root of our problem in the UK. Using mongrel bees and taking pot luck with the matings will result in a wide range of temperaments in the offspring. The only beekeeper who claims experience with the old English bee, and has committed his findings to paper, is Bro.Adam. He considered the bee had some very desirable features but also some very bad ones, notably bad temper. BIBBA claim that there are pockets of the old English bee with a good temper and campaign for its breeding and use in the UK.

Before leaving this introduction to understanding the temperament of our bees it must be pointed out that other characteristics will also be present and variable with mongrels such as high tendency to swarm, heavy propolis gathers, early starters/late finishers, longevity, economy on stores for overwintering, resistance to different diseases, etc. We know from experience that it possible to eliminate bad temper fairly quickly by culling those queens producing bad tempered bees. However, it is nigh impossible to control more than two variables without recourse to specialised isolation and mating techniques. Our own breeding programme concentrates on good temper and minimum swarming tendency which can be achieved by any beekeeper if they put their mind to it. Unfortunately, it is a very small minority who control the temperament of the bees they keep and we put this down to two traits that have developed over the last 50 years as follows:

a) Newcomers have been encouraged to wear very adequate protective clothing which has become available making them 'safe' from stings under most conditions.
b) Few associations encourage their membership to manipulate their bees without gloves.

3.1.2 Appearance, behaviour and characteristics.

Listed below are the known characteristics of the races that have been imported and which one day may be imported again now that UK is Varroa infested. The listing is in the following order; appearance, advantages, disadvantages. We consider that Bro.Adam's book 'In search of the best strains of bees' is essential reading on this topic.

French black bee. Very dark/black. Very good honey gatherer. Very aggressive, attacking (not defending) instinct is a real menace.
Swiss bee (*Nigra*). Pitch black. Honey gathering ability, industrious, wintering ability. Swarming, temper and aggressiveness.
Old English bee. Dark brown. Longevity, good wing power, industrious, good comb and cappings. Restricted fecundity, susceptible to disease, doubtful temper.
Heath bee (Dutch). Dark/black. Good vitality. Prolific swarmer.
Finnish bee. Dark/black. Found to be unsuitable for the British Isles.
Apis mellifera ligustica. (1) From Italian Alps; dark leathery. Industrious, resistant to Acarine, good temper. Drifting, more economical than its southern counterpart. (2) From S.Italy; bright yellow. Industrious in a flow, resistant to Acarine, good temper, will cross well with most races, excellent for amateurs. Not thrifty, honey into brood and breeds to excess at the end of a flow, bad robbers, lacks stamina, longevity and wing power, slow spring build up. This bee,

more than any other, has contributed to the advances in beekeeping over the last 100 years. It is interesting that strains developed in USA and New Zealand are very susceptible to Acarine. *Apis mellifera carnica.* Leathery/black version of Italian yellow bee. Good temper, industrious, resistant to disease, good stamina, minimum propolis, thrifty, good orientation, produces good crosses with Carniolan drones. Limited fecundity, excessive swarming, small colonies, bad comb builder.

Apis mellifera caucasica. Grey and hairy. Good temper, long tongue, stores in minimum of space. Limited fecundity, not hardy, excessive propolis and brace comb, susceptible to Acarine and nosema, not good for cross breeding.

Apis mellifera cecropia. Similar in appearance to the carnica. Good temper, high fecundity reluctance to swarm, exceptional thrift, good for cross breeding. Sensitive to inbreeding, excessive use of propolis and brace comb.

3.1.3 Other points.

• Older beekeepers will remember the days when it was possible to purchase Italian, Carniolan and Caucasian queens of pure race. Today it is now only the New Zealand strains that are available with their attendant disadvantages.

• The average beekeeper is interested only in good temper and low swarming propensity. This can be achieved by breeding from mongrels and culling the ones with adverse characteristics. We have found that a cull of about 10% of all queens reared from our 'best' stocks is required, some years more and some years less. It is unfortunate that more hobbyist beekeepers do not do follow this simple route.

• We consider that dispensing with gloves and rearing and culling the bad strains of queen essential objectives for all BBKA members. The advent of better protective clothing and the obsession for advising its use has, in our opinion, done much to harm the public's image of beekeeping.

3.2 The main external features of the drone and the two female castes.

3.2.1 Structure common to all three: head, thorax and abdomen:

- Head   2 compound eyes, 3 simple eyes (ocelli),
    2 antennae (scape + flagellum),
    2 mandibles,
    1 proboscis.

- Thorax - 4 segments (T1 to T3 plus A1),
    2 prs. wings (between tergites and pleurites of T2 & T3),
    3 prs. legs (on segments T1, T2 and T3).

- Abdomen - 6 visible segments (A2 to A7),
    3 invisible segments A8 to A10 (which are internal and part of the sting chamber).

3.2.2 Physical size: Worker about $5/_8$in (16mm) long.

	Queen about 1in (25mm) long but larger than a worker in diameter; nb. queen excluder.
	Drone about $^3/_4$in (19mm) long (much fatter than q. or w.).
3.2.3 Head:	Worker - triangular in shape with long proboscis.
	Queen - similar to worker but rounder with short proboscis.
	Drone - almost circular (n.b. large compound eyes). Antenna has extra joint (flagellum 12 segments) and mandibles are very small. The proboscis is short.
3.2.4 Thorax:	Worker/queen - similar in size; dorsal side in queen appears hairless cf. a worker.
	Drone - larger/stronger, larger wings (stronger flier).
3.2.5 Abdomen:	Queen - very distinctive (long/tapering).
	Drone - also distinctive (fat and furry).
	Worker - specialised (wax and Nasonov glands).
	All castes have 10 prs. spiracles on segments T2 to A8, the last being invisible and inside the abdomen.
3.2.6 Legs:	All have the same formation - coxa, trochanter, femur, tibia, 5 tarsal joints
	Note: the fore legs of all 3 castes have an antenna cleaner and only workers have pollen collecting equipment on the rear legs. The hairs on the basitarsi of the fore legs are used as a brush for cleaning the eyes and head in all castes.
3.2.7 Wings:	Forward pair large with fold to engage with the hamuli on the smaller rear wing. Drone wings much larger. In all 3 castes the wings are folded at rest and lay flat on dorsal side of the abdomen.
3.2.8 Hair:	The whole of the exoskeleton is covered in plumose hairs which have an important function in the worker for trapping pollen

3.3 The function of the hypopharyngeal glands, the Nasonov gland, the wax glands, the alimentary canal and the sting of the worker and queen.

The hypopharyngeal, Nasonov, wax and sting glands are all exocrine glands and all have secretions which are very important to maintain the social structure of the whole colony. Note that not all books regard the sting as an exocrine gland but the sting has glands which secrete externally. The secretions are the alarm pheromone, isopentyl acetate, in addition to the venom.

3.3.1 Hypopharyngeal: located above the pharynx and under the frons. There are 2 glands one on each side of the head. An axial duct from each gland opens separately via 2 small pores on to the hypopharyngeal plate and the secretion accumulates in the pre-oral cavity (food channel in the base of the proboscis inside the pharynx or mouth).

a) The secretion from the glands is called brood food or royal jelly. They are basically the same but there are chemical differences depending on whether it is being fed to worker or queen larvae in order to produce the two female castes. It should be noted that the adult queen lives solely on a diet of royal jelly which is very high in protein. The feeder bee folds back its proboscis, opens its mandibles and raises its labrum to allow the entry of the queen bee's proboscis.

b) The glands are large in the worker, rudimentary in the queen and non-existent in the drone. Glands are thread-like with round cells (acini) which are active in workers 5-15 days old (nurse bees) after which the acini shrink and become inactive (they become atrophied). It should be noted that older foraging bees can, if the needs of the colony so require it, consume pollen and re-activate their hypopharyngeal glands for brood feeding. Workers deposit the brood food in the cells, directing it into place with their mandibles.

c) In older bees the hypopharyngeal glands are believed to be the source of invertase although extracts from the glands have not been separated for analysis. Dade makes the point that analysis of the secretion of the postcerebral glands show surprisingly differing results, some workers indicating the presence of invertase, others denying it. For practical purposes modern thinking is 'invertase from the hypopharyngeal glands'.

d) Brood food (bee milk in some books) and royal jelly contains mainly:

- proteins,
- several vitamins of the B group,
- vitamins C & D but not E,
- (E)-10-hydroxy-2-decenoic acid (10-HDA), which acts as a preservative preventing bacterial infection of the food while it is lying in the cell, but note this comes from the mandibular glands.

3.3.2 Nasonov: (see diagram below) The scent gland located at the anterior end of tergite A7 in a transverse groove adjacent to the posterior end of tergite A6. It is exposed in use by flexing the two abdominal tergites to allow the secretion to evaporate; the process being assisted by fanning.

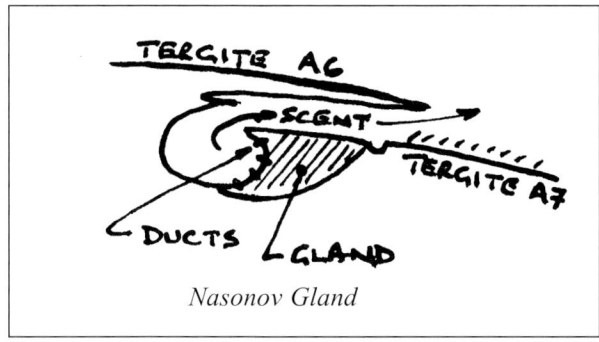

Nasonov Gland

a) The gland secretes into the scent canal (transverse groove) the secretion containing:
- E citral and geranic acid, the 2 most important components,
- Z citral, geraniol, nerol, nerolic acid and E,E farnesol.

b) Maximum production is found in foragers; it is low in winter and high in the spring.

c) The pheromone is highly attractive to bees and is in much evidence when bees are shaken, swarms being hived, etc. It is only found in workers and is used generally at or near the hive entrance. Colloquially known as "the come in and join us" pheromone.

d) Associated with the collection of water (foragers release Nasonov pheromone) but not with nectar or pollen. Also used on rich sources of sugar syrup as this also is odourless.

3.3.3 The sting. (See diagrams below) The actual sting consists basically of 3 parts: stylet and 2 lancets (barbed). There are other associated parts namely, bulb and umbrella valves, the rami, oblong plate (fixed), quadrate plate (moveable), triangular plate (moveable), muscles connected to the oblong and quadrate plates, acid sac, acid gland and alkaline gland.

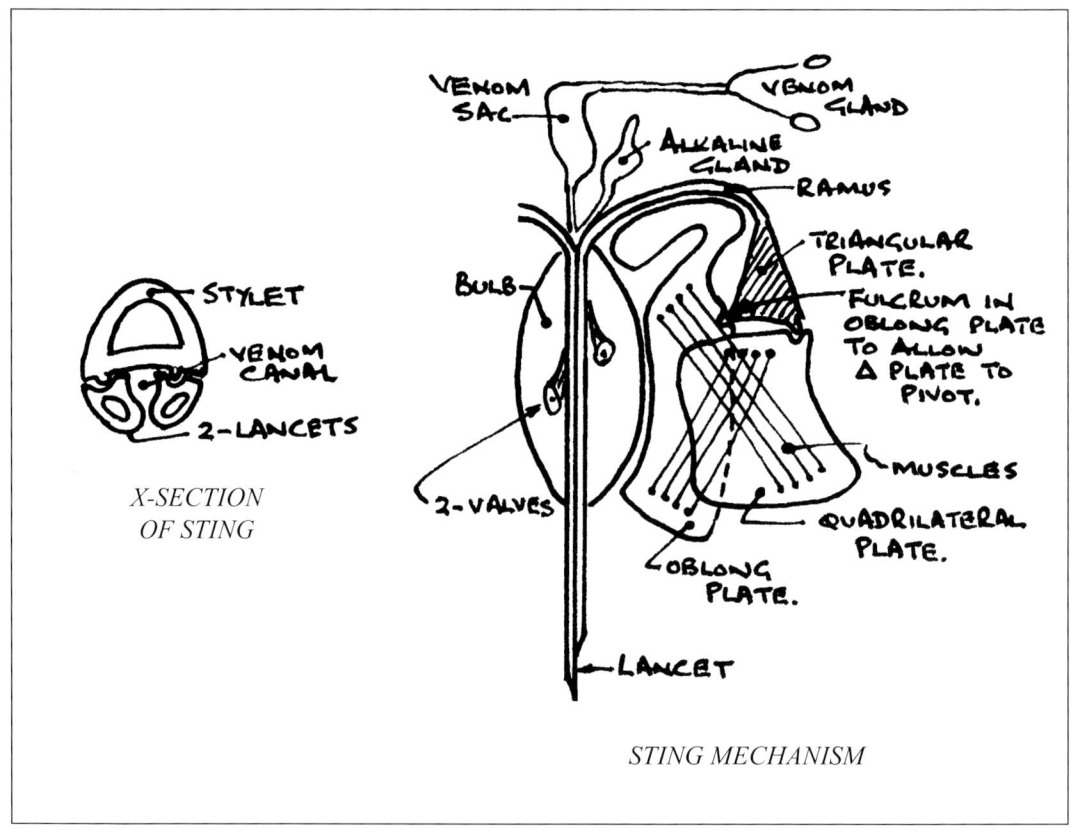

X-SECTION OF STING

STING MECHANISM

a) The lancets slide on tracks on the stylet and the 3 parts form a tube (the venom canal).

b) The stylet and lancets are connected to a bulb via vents and umbrella valves which deliver the venom from the bulb down through the venom canal.

c) Each lancet is connected to a corresponding ramus which in turn is connected to a pivoting triangular plate. Movement of the triangular plate is rotational, activated by a moving quadrilateral plate and also a fixed oblong plate. One corner of the triangular plate is pivoted to the fixed oblong plate.

d) As one lancet is pushed forward, a simultaneous action withdraws the other and vice versa.

e) Movement of the ramus and lancet also operates the umbrella valve in the bulb, thereby ejecting venom from the bulb into the venom canal.

f) The bulb (venom reservoir) is connected to the venom gland via the acid sac. Another gland, the alkaline gland appears to supply the bulb but actually secretes into the sting chamber.

g) The use of the alkaline gland is not exactly known. Suggested uses are lubricant for the stylet and in the case of the queen for gumming eggs to the base of the cell. Neither uses are proven. The acid gland produces the venom which contains:

Phospholipase A *	enzyme
Hyaluronidase *	enzyme
Acid phosphatase	enzyme
Allergen C *	
Mellitin *	
Mellitin F	
Mast cell degranulating peptide	
Secapin, tertiapin, etc.	
(* = main allergens).	

Note that the above are the main constituents of bee venom, but other substances are present in small quantities.

h) Use of the sting - only known to be used for self defence and defence of the colony. It should be noted that different strains of bee have a greater or lesser degree of colony defensive behaviour.

i) When a bee leaves its sting in the victim and tears itself away, the associated ganglion and muscles are still attached to the sting and the lancets continue to operate, the sting penetrating deeper and deeper. It should be scraped out with a knife or finger nail as quickly as possible. The enzymes in the venom cause the release of histamine from mast cells which causes swelling. Other allergens effect the nervous system.

j) The queen only uses her slightly curved sting against rival queens. Most books state that her sting is barbless; there are however 2 or 3 small barbs at the tip of the sting.

3.3.4 Wax glands and wax production (see diagram below):

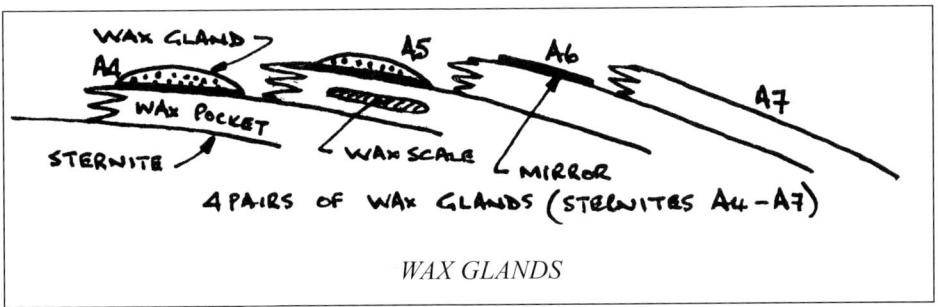

WAX GLANDS

119

The glands are located inside the exoskeleton on sternites A4-A7 inclusive. There are 2 glands on each sternite, making 4 pairs in all. The glands secrete a liquid which passes through the mirrors and oxidises as a flake of wax in the wax pockets. The glands, mirrors and pockets being known colloquially as the "waistcoat pockets".

a) Wax is secreted at relatively high temperatures (33°C-36°C) after consumption of large amounts of honey. Various estimates are quoted for the amount of honey to metabolise 1lb of wax. 5-8lb. being a realistic estimate.

b) Wax glands are best developed in worker bees 12-18 days old.

c) When building comb, bees hang in festoons near the building place, after gorging themselves with honey, waiting for the wax to form.

d) The wax glands inside the exoskeleton are covered with fat bodies and other cells. The major components of beeswax are:

- hydrocarbons	16%
- monohydric alcohols	31%
- hydroxy acids	13%
- fatty acids	31%
- diols	3%
- other substances	6%

e) The chemistry of how the wax is produced and how it diffuses through the mirrors is extremely complex and it is not necessary to know the detail; however it is necessary to know that a diffusion process is involved.

f) Wax is normally white but can be tinged with yellow hues caused by pigments that originate in pollen (eg. when a colony is working dandelion, new comb is noticeably coloured yellow).

g) For completeness a few of the physical properties of wax are:

- SG = 0.95 (Honey-like odour & faint taste),
- when pure, melts at 147.9°F (64.4°C), solidifies at 146.3°F (63.5°C).

3.3.5 The structure and function of the alimentary system.

The alimentary canal (transmitting food from mouth to anus) consists of :

Pharynx (mouth),
Oesophagus (tube),
Honey sac or honey stomach or crop,
Proventriculus (valve),
Ventriculus (mid gut),
Pyloric valve,
Small intestine (+Malpighian tubules),
Rectum,
Anus.

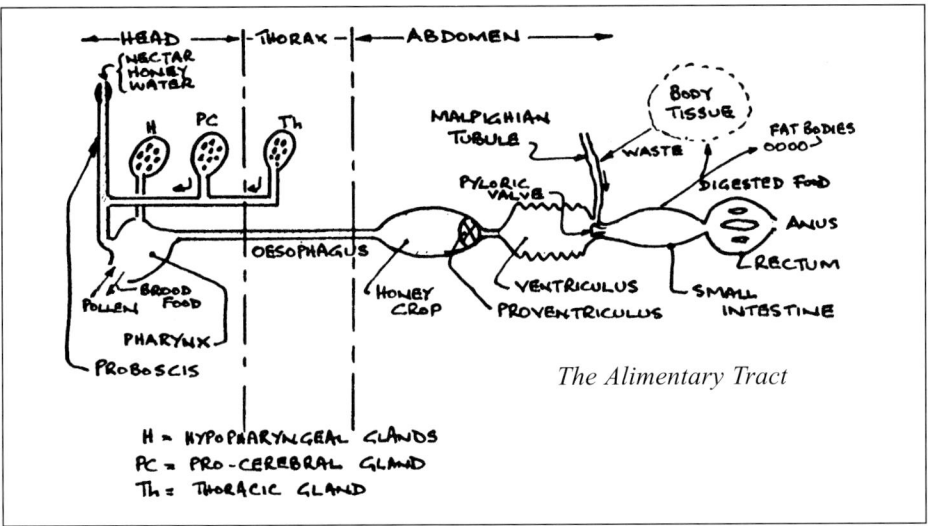

The Alimentary Tract

Note that the salivary glands are not strictly part of the alimentary canal but are often included in literature on the bee and its anatomy. A simple account would not include their use and may be ignored for this module. Each of the above are examined below; reference should be made to the diagram above of the alimentary tract.

3.3.5.1 Pharynx: the true mouth cavity with:

1) Inlets from the hypopharyngeal glands and the proboscis.
2) Entrance to the oesophagus.
3) Inlet/outlet via the mandibles.

3.3.5.2 Oesophagus: a tube through the thorax connecting pharynx to the honey sac. Its only function is to provide a passage for food in both directions.

3.3.5.3 Honey sac: transparent bag at the anterior end of the abdomen and capable of considerable dilation. Maximum load c. 100mg with an average load 20 to 30mg. Its function is to carry and temporarily store nectar, honey and water.

3.3.5.4 Proventriculus: is a one way valve which:-

1) Prevents (when necessary) nectar and honey flowing into the ventriculus.
2) Separates the pollen from the nectar/honey. The nectar is retained in the honey sac and the pollen is collected in four pouches behind the four lips of the valve; the filtration being done by very fine hairs (capable of filtering to 1μm). When a pouch is full, a bolus of pollen is passed into the ventriculus.

3.3.5.5 Ventriculus: is the true stomach of the bee where digestion of foods take place. Note the shape which is like 'gas mask tubing' which is ideally designed for peristalsis which moves the food

121

and waste matter through the final parts of the alimentary canal. The ventriculus is lined with an epithelium which produces proliferating cells continually; these sloughed off cells contain enzymes. The enzymes enter the pollen grains through the germ pores and digest the proteins in the grain. The resulting nutrients are then absorbed through the walls of the ventriculus and into the haemolymph (blood) in the abdominal cavity. A similar digestion process occurs with honey and nectar, the enzymes breaking down the polysaccharides into monosaccharides which can then be absorbed into the haemolymph through the walls of the ventriculus. The epithelium is covered with a jelly-like substance, the inner surface of which forms the peritrophic membrane. This encloses the food and is said to prevent abrasion of the ventricular wall. Waste matter produced in the ventriculus contains pollen husks, fat globules and uric acid and nitrogenous waste from the Malpighian tubules which enter the alimentary canal at the posterior end of the ventriculus.

3.3.5.6 Pyloric valve: this is a thickening of the walls at the anterior end of the small intestine just behind the entry point of the Malpighian tubules. The valve is lined with backward facing hairs said to assist in directing the contents in the backward direction.

3.3.5.7 Small intestine: is constructed in a fluted formation of 6 flutes. This type of formation provides a large surface area and slows down the passage of food. This suggests that absorption of digested food could take place in this region but as far as we are aware has not been proven.

3.3.5.8 Rectum: this is similar to the honey sac in so far as it can be dilated to such an extent that it can fill the whole abdomen. Six rectal pads are found on the outer surface of the rectum. They are quite distinctive but the function is so far unknown. It has been suggested that they extract water from the contents and return it to the haemolymph; the microscopic structure does not support this. Other suggestions are for the absorption of digested fats and maintaining the concentration of salt in the blood. The function of the rectum is to store waste products prior to the bee taking a cleansing flight.

3.3.5.9 Anus: the final outlet from the alimentary canal.

3.3.5.10 Malpighian tubules: there are about 100 of these tubules, closed at the distal end, which join the alimentary canal adjacent to the pyloric valve. They spread throughout the abdominal cavity and absorb, through their walls, nitrogenous waste from the blood. These are the main excretory organs and collect the waste matter from the nutrients used in the blood and tissues.

Note that two adult bee diseases are associated with the alimentary canal, these are:

>*Nosema apis* (Zander) - in the ventriculus restricting the digestion of proteins from the pollen grains.
>*Malpighamoeba mellificae* (Prell) - in the Malpighian tubules.

3.4 The factors in the production of brood, which result in workers, drones and queens.

The major factors which result in the production of the three honeybee castes can be summarised as:

a) Metamorphosis and the duration of the development stages of the larvae and the pupae.

b) Caste differentiation in the two female castes.

Each are examined in the following paragraphs.

3.4.1 Metamorphosis and the duration of the development stages of the larvae and the pupae.

3.4.1.1 Definitions:

• Metamorphosis - change in form by magic or by natural development or change (usually rapid) between immature form and adult state.

• Caste (zoological definition) - form of social insect having a particular function.

It should be noted that some books and authors recognises two female castes and the drone (male caste). We believe the question of two or three castes is a matter of opinion, there being arguments in favour of both. As three have been recognised for over 100 years, the change seems unnecessary and we have not changed our original notes.

3.4.1.2 The three castes:

• worker - from a fertilised egg (female)	32 chromosomes
• queen - from a fertilised egg (female)	32 chromosomes
• drone - from an unfertilised egg (male)	16 chromosomes

3.4.1.3 Stages in the life cycle:

	WORKER	QUEEN	DRONE
OPEN CELL:			
Egg	3d	3d	3d
Larva (4 moults)	5d	5d	7d
SEALED CELL:			
Larva/pro-pupa (1 moult)	3d	2d	4d
Pupa (1 moult)	10d	6d	10d
From egg to emergence	21d	16d	24d
AFTER EMERGENCE:			
Summer bee	6w	c.3y	c.4m
Winter bee	c.6m	ditto	n/a

Note that the above times can vary by a few hours before emergence due to variations in temperature of the brood nest. [c.= approximately, d = day, w = weeks, m = months and y = years].

3.4.1.4 Description of the stages in the life cycles.

Worker (before emergence):

1st day of egg	- vertical, stuck to the bottom of the cell and parallel to the cell walls
2nd day	- at an angle of c. 45°
3rd day	- horizontal, egg lying on the bottom of cell. Egg hatches after 3 days.
4th-8th day	- larva grows, moulting every 24 hours, until it fills the whole cell diameter. The cell is sealed on 8th day after the larva's last meal.
8th-21st day	- the connection between the ventriculus and the hind gut opens and the Malpighian tubules open into hind gut; excreta enters hind gut and is voided into cell (Important in the spread of EFB). Larva changes position and stretches out the full length of the cell (head outwards) and spins a cocoon. Metamorphosis occurs and the larva changes to a pupa after 5th moult 3 days after sealing. The pupa is still white but of adult form. It completes development, slowly changing colour and emerges from its cell by nibbling the capping on day 21. The 6th moult occurs just before emergence.

Queen and drone:

Similar but with the different timings shown above.

After eggs hatch, the larvae are immediately capable of eating food:

Workers and drones - are progressively fed with brood food for first 2-3 days, then a mixture of brood food, pollen and honey. No food is consumed during the pupal stage.
Queens - are fed by mass provisioning with royal jelly throughout the larval stage. Queens are generally over fed and excess can usually be seen in the cell after emergence.

3.4.2 Caste differentiation in the two female castes.

3.4.2.1 Caste determination. Both the queen and the worker are derived from a fertilised egg. Therefore, differentiation between queen and worker cannot be due to any genetic differences and must be due something else (e.g. feeding).

3.4.2.2 The queen is fed on royal jelly (a glandular secretion from the hypopharyngeal and mandibular glands of the worker bee) throughout the whole period from hatching of the egg to the propupa stage of development. A plentiful supply of royal jelly is available at all times for the larva. The larva continues to feed on the same diet after the cell is sealed.

3.4.2.3 The worker is fed on brood food for the first 2-3 days after the egg hatches, then honey and pollen are added to the brood food up to the sealing of the cell. Note that no food is left in the cell as in the case of the queen.

3.4.2.4 Analysis of larval food has given variable results.

124

- In 1888 Planta postulated that differing foods determined the female caste (queen or worker),
- In 1943 Haydak postulated that it was not the type of food that determined caste but the amount consumed. Experiments to test this theory have not been fully conclusive.
- In 1956 Weaver carried out feeding experiments to confirm Von Rheim's work in 1933 that there was a 'fugitive' substance in the larval food.
- In 1961 Jay disputed the interpretation of Weaver's results.

It is clear that, at present, the exact mechanism determining caste of the female is not fully understood and more work is required to answer the problem fully. However for the purposes of this examination, the present thinking about caste determination may be summarised as follows:

- It is triggered by the sugar content of the larval diet under 3 days old,
- The sugars induce increased food intake and growth of the corpora allata (CA), an internal gland which produces a special hormone,
- The hormone in some way is responsible for the differentiation in the anatomical and morphological characteristics. The precise mechanism is unknown at present.

3.4.2.5 Cell differences. The only other differences are in the cell (ie. worker and queen cells); the size and orientation have been shown to have no effect on caste determination.

The problem of caste determination is not an easy one to understand and this complicated subject is considered further in the Orange Book, "Beekeeping Study Notes (Volume 2)" for Module 5, for those who may be interested.

3.5 The mating of drones and queen.

3.5.1 General.

- Queen mates on the wing between 5-20 days after emergence.
- If she has not mated in c.3 weeks she is no longer capable of mating properly and is known to be 'stale'. Sperm cannot migrate through the duct leading to the spermatheca.
- Bees in a colony with a virgin queen become more and more aggressive to her until she mates.
- This aggressive behaviour may be responsible for driving out the virgin queen before she is too old to mate and becomes stale.
- After mating, the bees are very attentive to the queen, grooming her and forming a court around her.

3.5.2 Drones.

- Drones often have collecting points where they tend to congregate. Such a congregation point attracts drones from a wide area ensuring drones of varying strains and thus minimising inbreeding. It has been observed that mating only occurs at heights of greater than 10m (33ft) and less than 40m (130ft) above the ground. Also, the height of mating is inversely proportional to wind speed.
- There are drone congregation areas (DCA) where drones collect to mate. How these are chosen is unknown but some sites are used year after year.

• Drones in the area release a pheromone from their mandibular glands which not only attracts queens but also attracts other drones into the area ensuring a good mix of drones and genetic material from other colonies and other apiaries in the area. When many drones have congregated the pheromone density is high ensuring a good attractant to queens and the recruitment of other drones.
• When a queen enters the DCA the drones locate the queen initially by sight and then by sex attractant pheromone. They follow the queen in a comet tail formation of as many as 100 drones until one drone catches the queen and mating occurs.

3.5.3 The mating.

We have only been able to locate one description of the actual mating by Köeniger 1984 and the following is our summary of a much fuller account.

• Queen flies to the level of the drones and locates the DCA by the drone pheromones.
• Once in the DCA the queen responds to the drone pheromone by holding the entrance to her sting chamber open whilst in flight.
• The strongest flying drone from the 'comet' reaches the queen and grasps her with his legs. The fore and middle legs on the dorsal side of her abdomen and the hinds legs on the ventral side of her abdomen.
• The drone then bends his abdomen and everts his endophallus which enters the sting chamber. The drone then becomes 'paralysed' and loses his hold and swings backwards.
• In this position he is carried along by the queen. The endophallus being shaped to fit in the sting chamber making it impossible to be released while the endophallus is everted.
• The cervix of the drone's penis enters the vagina and after a short pause eversion continues (even in the drone's paralysed state) and semen is ejaculated into the median and lateral oviducts.
• The pressure of the semen into the oviducts also pushes out part of the endophallus which becomes detached from the bulb.
• The drone drops to the ground to die and part of the bulb remains in the sting chamber. Some mucus from the mucus glands of the drone coagulates and forms a sealed plug.
• A second drone will mate with the queen in exactly the same way except that the second drone removes the mating sign first with his own endophallus.
• Perhaps up to 5 matings occur until the oviducts and vagina are full of sperm. This then enters the spermatheca by contraction of the oviduct muscles of the queen.
• Further matings (5-15) occur on 2 or 3 separate mating flights.
• Mating continues until the spermatheca is full of sperm. After mating the queen has sufficient sperm to last her life of 3-5 years.
• The vaginal opening of queens returning to the hive after mating flights often contain the male genitalia which are removed by the bees inside the hive.
•The queen is unable to mate with drones of other species due to the unique structure of the endophallus and the sting chamber of the queen allowing the interlock to take place and the queen being unable to release the drone.

3.5.4 Mating flights.

These normally occur in good weather when there are plenty of drones flying (say noon to 4 pm) at temperatures 20°C and greater. High winds discourage mating flights. Average length of time of mating flights have been observed to be c.20 mins. in April which decreases to c. 12 mins. in June.

3.5.5 The mated queen starts to lay c. 2-4 days after mating is complete. Egg laying is often erratic when the queen starts to lay and more than one egg per cell sometimes occurs. This phenomenon generally soon disappears.

3.6 The main stages in the development of the brood from egg to emerging adult and also the life expectancy of workers, drones and queens.

See Section 3.4 above.

3.7 The changing circumstances throughout a year that influences the egg laying of a queen, indicating how the numbers will vary.

3.7.1 General.

The egg laying rate of the queen will be directly represented by the total population of the colony. Reference to Appendix 8 shows the annual population cycle of a normal disease free colony which doesn't swarm during the season. The peak population is shown as c. 60,000 at the end of June. This is a massive colony. It is now necessary to put some numbers to the egg laying rate of the queen.

3.7.2 Colony size/egg laying rate of the queen.

The productive colony should have a maximum sized brood nest during the first three weeks in June, which will produce the maximum foraging force at the beginning of July. The number of adult bees in the colony will be about double the amount of brood and the amount of brood will be approximately 20 times the daily egg laying rate of the queen. Putting some numbers to this, we have:

Eggs per day	Total brood	Adult bees
1000	20,000	40,000
1500*	30,000	60,000
2000	40,000	80,000!

With this amount of brood, how many BS frames does this represent? The answer to this, of course, depends on the percentage fill in each frame which will not only contain brood but stores of honey and pollen. Again putting some figures to this on the basis of 5000 cells/BS frame, we have:

Stores	Brood	Brood/frame	20k brood	30k brood
25%	75%	3750	5 frames	8 frames
40%	60%*	3000	7 frames	10 frames
50%	50%	2500	8 frames	12 frames

With a reasonably prolific queen and using a 60%* brood/frame (which from experience seems to be about right), then 10 frames of brood will be required in the productive colony. This could be used as a target to achieve in the build up; it will be a massive colony with 60,000 bees. If the brood chamber is full of brood and a flow starts, then the only place nectar and honey can be stored is in the supers, just where it is required.

Colony size is an interesting concept and writers seem to vie with every other to give a higher number (rather like fishing stories!). Dr. Jeffree, in some of his experiments, measured 510 colonies and the biggest, a really large one in his own words, contained 47,700 bees. So the estimate above is a bit on the large size, but it is interesting to try and quantify it in terms of frames of brood.

Queens do not lay well when there is no income to the colony. Therefore, if there is no flow the colony will require to be fed in order to build it up. A further factor in the equation is whether foundation is to be given or drawn comb; to draw foundation requires a flow or feeding. A flow requires good weather.

We now have all the variables that influence the egg laying of the queen, namely;

    a) the fecundity of the queen,
    b) the weather,
    c) flow (depends on forage available) or feeding,
    d) foundation or comb.

The circumstances which change are the weather which is beyond the control of the bees or the beekeeper.

3.7.3 Additional information.

The weight of an egg laid by a queen = 0.13mg and the weight of a queen in full lay = 130mg which is considerably more than a worker honeybee which weighs 90mg.

Therefore if a queen is laying 1000 eggs per day this is equivalent to a weight of 130mg (the weight of the queen herself). This means that the queen has to be fed continually night and day in order to metabolise the food to produce this number of eggs. There are references in the literature of queens laying 3000 eggs per day at the peak of the season, claims which we find difficulty in comprehending.

1000 eggs per day = 1 egg laid every 1.5 minutes which is a phenomenal feat when feeding and grooming is taken into account.

3.8 The nutritional requirements of honeybees and their main sources.

3.8.1 General.

• Nutrition in the animal kingdom is a chemical process changing carbohydrates, fats, proteins, etc. into bodily materials (muscles, tissues, etc.) and also into energy, both mechanical and heat.
• Honey/nectar (carbohydrates) is broken down into simple sugars and then converted into fat and glycogen during the nutritional process.
• Pollen (protein) is split into basic building blocks called amino acids and fats which are either absorbed or broken down to glycerol or fatty acids.
• Honey is the energy source and pollen is for growth, repair and development.
• Little is known about the digestion of fats and vitamin requirements of the honeybee.
• Most of the research work on nutrition in connection with honeybees was undertaken a long time ago and very little work on the subject has been undertaken during the last 20 to 30 years.

3.8.2 The value of honey to the colony.

• The major carbohydrates in nectar or honey are sucrose, glucose and fructose. In addition, honey contains many other sugars (trisaccharides and higher orders) but only 5 are understood to be sweet and nutritious to the bee (Von Frisch, 1934). All the higher order sugars represent a very small percentage of the total sugars and from a value to the bee aspect, they may be ignored.
• Energy required during flight is derived exclusively from the breakdown of carbohydrates. The blood sugar content is very important and the following values should be noted:

Flying worker bee has	blood sugar level	c. 3%
Unable to fly if	blood sugar level	1%
Motionless if	blood sugar level	0.5%

Flying bees require c. 10mg per hour and drones c. 30mg per hour.
• Ambient temperature is very important, Olaerts in 1956 found with caged bees :

at c. 52°F (11°C) a bee requires c. 10mg sugar per hour
at c. 98°F (37°C) a bee requires c. 0.7mg ditto
at c. 118°F (48°C) a bee requires c. 1.4mg ditto

• Hypopharyngeal gland development is not initiated on nectar/honey only, pollen is a necessary part of the diet.
• Beeswax is produced by the metabolism of sugars by the fat bodies and wax glands.
• The chitin of the exoskeleton is a nitrogenous polysaccharide made up of glucose molecules in combination.
• The amount of honey required to rear one worker bee has been estimated to be c. 100mg in the range 50 to 150mg.
• Water is essential to dilute the sugars to a 50:50 ratio in order that the bee may metabolise them. It should be noted that the bees within the hive continually exchange food.
• The nectaries of flowering plants and trees are the main source for the nectar gatherers from the hive.
• Once the nectar is converted into honey, the cells are sealed with wax. This preserves the honey for use by the bees when there is no fresh nectar available eg. winter months.

3.8.3 The value of pollen to the colony.

Pollen is the male germ cell of flowering plants (angiosperms). It is collected from the stamens of flowering plants by foraging bees. It has two major uses:

a) It is the principal source of protein, fat, vitamins and minerals in the honeybee diet.
b) It can provide a surplus product from the apiary.

In this section of the syllabus we are only concerned with (a) above.

Pollen demand in the colony is related to the amount of unsealed brood. Bees cannot rear brood without pollen because the nurse bees would not be able to produce brood food from the hypopharyngeal glands. A strong colony will collect c. 50 - 100lb during a season.

It requires 70 - 150mg of pollen to rear one adult bee.
About 200,000 bees are reared during a season thus accounting for more than 50% of the income.
The balance is used by the adult bees preparing for winter (increasing their fat bodies) and/or stored in the comb for use early the following year before new supplies become available.

Note the weight of a worker bee = c.90mg and that 1lb of bees contains c. 5000 bees.

Pollen is rich in protein and it is essential as a body building material for growth/development and for the repair of worn out tissue. It also has the very important function of stimulating the development of the hypopharyngeal glands and the fat bodies of the winter bee. The protein content varies between different pollen types and also from flower to flower in the same foraging area. A protein content of c. 35% is typical of a high protein pollen eg. beans. Bees can discriminate between pollens by colour and odour; they cannot distinguish between the quality (protein content) of various pollens.

Pollen contains:
- proteins   7 - 35%
- lipids (fats/oils)   1 - 14%
- amino acids   ?
- carbohydrates   ?
- minerals   1 - 5%
- vitamins   ?
- enzymes   ?
- water   7 - 15%
- sugars   25 - 48%

There are wide variations in the content of different pollens and the bee more than likely receives a balanced diet due the variety of pollens collected and used.

The use of pollen for brood rearing and development of the adult bee:

1. Worker larvae are fed brood food only from 0 to 3 days and then on 4th and 5th day with pollen, honey and brood food.
2. Queens both adult and larvae are fed exclusively on royal jelly.

3. After emergence of the worker bee, pollen is essential for it to reach maturity in a healthy state. It depends on pollen for its orderly development of its glandular system while it is a house bee.

4. The pollen required to rear one bee = c. 120mg. If 200,000 bees are produced in the course of one year, then c. 50lb of pollen is required to be collected.

5. Pollen is also required for the glandular development of the newly emerged bee. First the hypopharyngeal glands and finally the sting glands just before it takes up guard duties prior to foraging.

6. The conversion of pollen into protein has been calculated to have an efficiency of about 40%, ie. 10mg of pollen will produce about 4mg of protein.

7. A strong colony at the peak of its brood rearing requires about ½lb pollen per day.

In areas where natural pollen is in short supply, particularly in the spring, pollen patties can be fed to colonies to stimulate spring build up. Pollen shortage often occurs where colonies are foraging on honeydew in pine woods.

3.8.4 Other points in connection with honey and pollen.

There are other important items in connection with nutrition which should be noted. These are:

Lipids - little is known in relation to the honeybee.

Vitamins - pollen is known to have a high vitamin content but again the effects of the various vitamins are largely unknown and little work has been undertaken. Vitamins are essential for the growth and development of living organisms.

Minerals - much the same situation exists with minerals as with vitamins but to a worse degree. The study of minerals in the nutrition of the honeybee and insects in general is probably the most neglected area.

Water - is essential for life in the animal and plant world.

3.9 The signs in a colony of a drone laying queen and laying workers. Explain how these may arise and how they may be dealt with.

3.9.1 Detection of a drone laying queen & the causes for this failure.

3.9.1.1 The visual signs:

• Unmistakable worker cells with drone cappings (raised).
• Presence of a queen (actually seen).
• Drones produced are small and abnormal (stunted).

3.9.1.2 During the season:

• Queen produces small areas of drone brood in the middle of large patches of worker brood.
• As the season progresses, worker brood becomes less and drone brood increases.
• Because some worker brood remains, it is clear that a queen must be laying.

- Eventually there will be nothing but drone cappings. At this stage the colony will be reasonably large.
- Drones are smaller and the abdomen stunted.

3.9.1.3 In the spring:

- Very difficult to detect at the first examination of the colony ie. a small colony with one or two frames of drone brood only (no worker).
- Is it a drone laying queen or laying workers?
- If a queen (drone layer), the laying pattern will be orderly i.e. compact patches of brood with very few empty cells.

3.9.1.4 Possible causes for a queen becoming a drone layer:

- Shortage of sperm - inadequate mating or due to age.
- Physical inability of queen to fertilise eggs correctly.
- Genetic fault.

3.9.1.5 Treatment:

- Requeen or
- Unite after removing old drone laying queen.

3.9.2 Detection of laying workers and description of why they occur.

3.9.2.1 Detection of laying workers.

- Drones in worker cells (typical raised domes).
- Drones produced in this way are small and abnormal (stunted).
- Laying pattern is scattered and haphazard (cf. drone laying queen which is compact and orderly).
- Colony endeavours to build charged queen cells (note: this can happen with drone laying queen but is unusual).
- Workers generally lay more than one egg/cell and they are more often than not on the sides of the cells because a worker's abdomen is much shorter than a queen. We have seen as many as 7 eggs in one cell but 2 or 3 are more normal for this condition.

3.9.2.2 Pheromones.

In the absence of the queen and normal worker brood there is an absence of pheromones from the queen herself and from worker brood. These pheromones, particularly that produced by the queen, inhibit development of a workers' ovaries. The workers start laying after about 21 days of being in the queenless state when their vestigial ovaries are sufficiently developed. Worker brood pheromone elicits pollen foraging.

3.9.2.3 Causes of a colony having laying workers:

• Queenlessness.
• Inability to produce emergency queen cells (no fertilised eggs).

Both the above can usually be attributed to the inability of the beekeeper to recognise the shortcomings of the colony for a period of up to 3 weeks.

3.9.2.4 Treatment.

It is generally agreed that little can be done except to shake the colony out near a strong stock and let them take 'pot luck' after the hive and all its parts have been removed, always providing the bees are disease free.

The following points are pertinent:

• Difficult (impossible?) to requeen; a colony usually kills an introduced queen.
• Bees are mostly old and not much use to another colony.
• If they are united to a queenright colony it has been found that there is the likelihood of them killing the queen of the colony to which they are united.
• Experiments conducted in France in 1989 on the introduction of queens to colonies of laying workers by dipping the queen in royal jelly and water (70% and 30% respectively) are claimed to be a successful treatment. We regard this to be of academic interest only and has no practical value in beekeeping management.

3.10 The seasonal variation in the hive population during a year including survival behaviour in winter.

3.10.1 The seasonal variation in hive population during a year.

See Appendix 8 for a full explanation as required in this section of the syllabus.

3.10.2 The survival behaviour of the honeybee in winter.

3.10.2.1 General.

During winter months there is no income, temperatures are low and the colony has evolved over millions of years to survive this period with minimum food and heat conservation by clustering. The metabolic rate of each individual bee is extremely low under these conditions. During this period there is no brood in the cluster and the inside cluster temperature ranges from 20°C to 30°C (68°F to 86°F) much too low for brood rearing. This is the basic simple answer but we believe that more detail is required for this examination and is contained in the paragraphs that follow in this section.

3.10.2.2 The origin of thermoregulation in the honeybee.

It was shown earlier that the origin of the honeybee, *Apis mellifera,* was in Africa and that during its evolutionary period it successfully migrated to northern latitudes. In so doing, it developed a way of flying in cold weather, keeping sufficiently warm in winter to survive and collecting enough stores until they became available again the following year. The origin of its ability to keep warm has developed from being able to fly at low temperatures.

• The heat losses from the exoskeleton are generally low and the honeybee is poikilothermic (takes up the temperature of its surroundings, ie. the ambient temperature).
• Its thoracic temperature (T_{Th}) is about 10°C (60°F) above the ambient temperature (T_{Amb}) when not flying.
• In order to fly, T_{Th} must be greater than 27°C (80°F) because at lower temperatures the minimum critical wing beat frequency cannot be generated. ie. the minimum frequency for take off (to get lift).
• This puts a constraint on the enzymes working in the flight muscles which will work at high T but not at low T. This is a characteristic of most enzymes, eg. fairly high temperatures are required for the enzymes in the nectaries of a flower to work in order to secrete nectar.
• The honeybee has developed a method of warming up these flight muscles prior to take off by operating simultaneously the wing elevator and depressor muscles in opposition to one another creating no movement but using energy and generating heat until the required critical temperature is reached ($T_{Th} = 27$°C).

It is this ability to generate heat with no apparent movement that provides the means for nest warming in cold climates not only in the winter but in the temperate summer temperatures of northern latitudes.

Honey flows in the tropics, the origin of the honeybee, are two per year and generally coincide with the monsoons. In the subtropics (eg. the Mediterranean) the honey flow is more or less continuous throughout the year which is why the yellow bee with its roots in southern Italy continues to rear brood into early winter and uses up its stores in the process if kept in northern latitudes. In northern latitudes there is generally only one main flow. The honeybee which has evolved for these climatic conditions builds up on the spring flow and then stores a lot of honey in a short time on the main flow and reduces its brood production when the flow comes to an end, thereby conserving its stores for winter use. It is to be noted that the production of oilseed rape is tending to distort this pattern, to some extent affecting the seasonal colony development.

The last feature that enables the honeybee to survive the cold winters in high latitudes is its ability to vary its age and instead of having a life span of 6 weeks in the summer this is extended to 6 months in the winter, ie. the summer and winter bees. In common with many other animals the honeybee develops fat bodies in large quantities in preparation for winter. Additionally, the hypopharyngeal glands are highly developed in c. 80% of the winter bees as compared with c. 20% of the summer bees. There is a marked variation in the content of the fat bodies as follows:

WINTER BEE   high protein, low fat
SUMMER BEE   low protein, high fat

A further feature of the ability to generate heat is for the incubation of brood. The brood nest temperature has to be maintained between very close limits with a mean of 90°F (33°C). If the brood nest temperature drops, even a few degrees, to 85°F or 86°F deformed bees will result with stunted abdomens and deformed wings. Very little research work has been done on the aspects of low temperatures in relation to disease but there are indications that low temperatures could be the trigger for the development of disease in the colony (eg. *Ascosphaera apis* spores causing chalk brood need a low temperature to germinate).

3.10.2.3 How the colony prepares for winter.

   a) It collects a summer/autumn harvest.
   b) Drones are evicted usually after the main flow in August/September.
   c) Dramatic loss in the summer bees.
   d) Large increase in the winter bee population resulting in:
       - Consumption of large quantities of pollen and honey.
       - An increase in size of the hypopharyngeal glands.
       - A large increase in the number of fat bodies which store glycogen (carbohydrate from honey), protein and fat.
       - Longer life as a result of no work to do rearing brood (lower CO_2) or foraging.
       - Life of bee pollen consumed ÷ (brood reared × CO_2 levels).

3.10.2.4 How the colony over-winters.

Winter is characterised by the absence of flowers and the consequential lack of pollen and nectar for the bee to forage. The days decrease in length and the ambient temperature decreases. The reaction of the honeybee is as follows:

• The queen stops laying, not abruptly but slowly reducing to nil (c. October/November) and starts again very slowly at first (c. January/February).
• The honeybee is poikilothermic, is immobilised at 45°F (7.2°C) and starts to die at 40°F (4.4°C). The only way it can survive is by clustering. The characteristics of the cluster are as follows:

   a) The cluster starts to form at 57°F (14°C) always below the food stored.
   b) The outer shell is 1 inch to 3 inches thick with a partially filled centre of festooning bees.
   c) The outermost bees forming the shell enter the empty cells.
   d) From 57°F to 45°F (14°C to 7°C) there is little change in the size of the cluster, the heat losses are regulated by heat produced by the bees.
   e) 45°F (7.2°C) is the critical temperature (T_{cr}) and if the temperature of the bees on the outside of the cluster dropped lower they would become immobilised and drop off. Below T_{cr} the cluster contracts and above T_{cr} the cluster expands.
   f) As the temperatures continues to fall the bees on the outside of the cluster bury their heads into the cluster and spread their wings forming an efficient heat protecting layer. Heat is generated inside the cluster to maintain the surface temperature at no lower than 45°F (7.2°C).
   g) Considering the heat losses from the cluster, clearly heat lost = heat generated by the bees.

The heat lost can be by conduction, convection and radiation. Conduction loss is virtually = zero; wooden frames and wax comb are poor conductors of heat. Convection losses due to air currents around the cluster are approximately = to the radiation (mainly infra red) losses. It follows then that the smaller the cluster (minimum surface area) the smaller will be the convection and radiation losses.

h) Consider the surface area of a bee = $2cm^2$. If there are 15,000 bees their total surface area = $30,000cm^2$. If they now form a cluster 18cm radius the sphere has a total surface area of $1000cms^2$. This represents a saving of × 30 in surface area and a corresponding saving in heat loss. If the cluster contracted further to 12cm radius the saving factor = × 66; a very efficient method of conserving heat and minimising energy used.

i) It is essential that the cluster maintains contact with the food reserves throughout a prolonged cold spell. It sometimes happens, but not very often, that a colony does get separated and starves after it has drawn on the individual fat body reserves of each bee.

j) It is to be noted that there is usually a connective cluster of a few hundred bees making contact with the food reserves and said to direct the main cluster around the combs during the clustering process. We have not been able to locate much meaningful information about this aspect of wintering behaviour.

k) As the days start to lengthen after the winter solstice (December 21st) the queen starts to lay again. We believe that the lengthening days are the trigger to start feeding the queen again and bring her back into lay. This behavioural aspect has not been proven and again there is a paucity of information about it in the classical literature.

l) By clustering and generating heat a normal healthy colony of *Apis mellifera* from northern latitudes with adequate stores can survive temperatures down to 35°C without any special management by man.

3.10.2.5 Temperature control.

• Measurements on over-wintering colonies have established that the internal temperature of the cluster varies quite widely between 68°F and 86°F (20°C and 30°C) providing there is no brood present.
• When brood rearing starts the brood area is maintained at the incubation temperature of 93°F (34°C).
• The surface temperature of the cluster is never allowed to fall below 45°F (7°C) the temperature that the bee becomes immobile.
• There is no evidence, as far as we know, to suggest that the bees on the outside of the cluster are continually changing and moving inside to warm up; it would be physically difficult for them to do this as they have their wings spread. However, it has been observed that a few bees move from the inside to the outer surface. This behaviour presumably is because of overheating in the middle of the cluster and the life of a bee is shortened when subjected to high temperatures for any length of time.

3.10.2.6 Ventilation requirements.

Food consumption during the winter produces energy and two by-products namely water, H_2O, contained in water vapour and CO_2. The process is known as respiration and more accurately referred to as tissue respiration to distinguish it from breathing. In its simplest terms tissue respiration is a chemical process converting sugar (glucose) and fats into other substances (CO_2 and H_2O) and releasing energy in the process. Enzymes act as catalysts for the various reactions that take place. Note that any one reaction will involve only one enzyme.

• The heat produced by the rapid micro movement of the thorax flight muscles in opposition to each other produces all the energy in heat with no movement of the bees' wings.
• The amount of water produced is approximately 4½ gallons during the course of winter. This has been derived in section 2.18.7.
• Any water vapour that condenses in the hive and evaporates will lower the temperature due to the latent heat of evaporisation thereby increasing the energy requirement from the colony.

It will be clear from the above that good ventilation is essential to remove the water vapour quickly and efficiently if the colony is to winter successfully. How this is achieved is still the subject of debate for practical beekeeping management; ie. whether to have through ventilation from bottom to top or lower ventilation only by means of an open screen below the cluster.

3.10.2.7 Humidity control.

The relative humidity (RH) of the brood nest is maintained at about 45% in order to keep the air moist and prevent the drying of the brood food. In summer the humidity rises to quite high values during the ripening of nectar and this is controlled by fanning. The regulation of temperature is considered to have a stabilising effect on the RH and the normal nest RH is always fairly high. The colony in winter has a less demanding task when there is no brood present and the only evidence to show that RH is being controlled in the winter cluster appears to be by temperature control, viz. the temperature ranging from 68°F to 86°F (20°C to 30°C) when there is no brood present. Note that a higher air temperature contains a higher percentage of water.

3.10.2.8 Other points on wintering.

• The colony requires to obtain water to dilute stored honey during the winter. There are two ways that this can be achieved without the bees venturing forth outside the hive and leaving the cluster. Firstly, honey is uncapped and because it is hygroscopic it will absorb water naturally and become diluted to the correct value; this is common place during wintering. Secondly, some water vapour condenses on the cold inner surfaces of the hive and is collected by the bees close to the cluster. Reabsorption of water from the rectum and small intestine has not been proven in the honeybee; the rectal pads were at one time thought to undertake this function but no proof has been forthcoming.
• Experiments have shown that the colony temperature increases by about 10° C if it is disturbed during the winter. This behaviour pattern is understood to be a defensive mechanism. The bees on the outside of the cluster are incapable of flying and they extend their stings; this will produce the alarm pheromone. The temperature rise is to warm the flight muscles ready for flying in defence of the colony. High temperatures shorten the life of a bee; it is for this reason that it is undesirable to disturb a colony in winter.
• It is possible to over-winter an observation hive on 2 BS frames if the hive is kept in a constant

temperature of about 60°F. Our own beekeeping branch maintained such an observation hive in the Natural History Section of the Plymouth Museum on a continual basis as a service to the city.

3.11 The effect of weather on a summer colony and foraging.

3.11.1 Introduction.

Again we have had difficulty interpreting the intent of the Examination Board in respect of this syllabus item. Reading it literally, the Candidate is required to be able to describe the effect of the weather on:
a) a summer colony and
b) foraging (presumably of the same summer colony).
But what is a summer colony? We have had this problem in our previous notes where the BBKA Examination Board use the word spring and summer yet, in their wisdom, fail to provide a definition.

3.11.1.1 Definition of spring.

The dictionary definition of spring is as follows:

"The season in which vegetation begins to appear, first season of the year, in the northern hemisphere March - May (Astronomical, from vernal equinox to summer solstice)."

The orbit of the earth around the sun shows an equinox in March (21st) and one in September (23rd) when the day and the night are equal (hence equinox). At these dates the declination of the sun is zero ie. at the first point of Aries and Libra respectively. Between these two dates the summer solstice occurs on about 21st June and the winter solstice on about 22nd December, the times of the year when the earth is furthest away from the sun.

So what does spring mean to the beekeeper? We consider that in the south of England it extends from the first inspection usually earlier than the equinox to the end of June just before the main flow is about to start and a little later than the summer solstice. The whole of spring, in our opinion, is concerned with managing the colony in order to maximise the population to take advantage of the main flow which generally starts in early July.

This is where the difficulty of definition starts to come in. The summer solstice is known as mid-summer day implying that summer started earlier and again by definition, summer follows spring. The further north one goes in the UK the flowering vegetation is later by approximately 2 to 3 weeks in Scotland and the north of England. This convinces us even more that the definition should be from first inspection to the main flow starting which is dependent on weather and latitude.

3.11.1.2 Definition of summer.

The dictionary definition of summer is as follows:

"The warmest season of the year, in the northern hemisphere from June to August. The astronomical definition is from the summer solstice to the autumnal equinox."

As we used the definition of spring finishing on the last day of June or just after the summer solstice, we believe that the period we should be considering in this part of the syllabus is from early July to the end of September. This is the period when generally the colony size is decreasing from its peak value just before the main flow.

Therefore a summer colony is one that has reached its peak and is starting to decline in size during the main honey flow in July up to the end of September. We will now address the effects of weather firstly on the colony itself and secondly on the foraging of that colony.

3.11.2 The effect of weather on the summer colony and its foraging activities.
During the 3 month period the normal activity will be as follows:
• The main honey flow will occur during July and by the end of this month the colony will have gathered all the stores it requires for the winter period.
• The flow will end abruptly and during August the colony will be arranging and ripening its stores for winter.
• During September the colony will be gathering pollen and using this to produce winter bees, the excess of pollen being stored and pickled for use in the following spring.
• Finally it will be gathering propolis and sealing its winter quarters against the forthcoming winter elements.

The amount of forage collected during July will depend upon the weather and the temperature. Warmth is required in order for the enzymes to function allowing the nectaries to secrete in profusion. Damp cold weather will have a very adverse effect on the amount collected. Cold weather in August will put an addition load on the bees, making the ripening process more energy consuming for the colony. Finally poor weather in September will mar the brood rearing and late autumn flow and associated pollen gathering. In short, the colony's chance of survival during the forthcoming winter will be reduced unless corrective action is taken by the beekeeper. Due note should be taken of the 'June gap' when often there is a dearth of forage just prior to the main flow.

3.12 The type of work done by a worker honeybee throughout its life including reference to summer and winter bees.

3.12.1 Description of the work undertaken by worker bees.

The work undertaken by the worker bee is generally dependent on the age of the bee and the development of various glands:

- 0 - 3d   House (hive) cleaning, eg. cells for queen to lay in.
- 3 - 9d   Feeding larvae/nursing.   *

• 9 - 18d	Ripening honey.	*
	Wax making/comb building.	*
	Ventilation/evaporation.	
	Temperature control.	
• 18 - 21d	Guarding/defence.	*
• 3 - 6w	Foraging.	

d = day, w = week and * = these activities require a glandular activity in the bee; 1st the hypopharyngeal and mandibular glands develop for producing brood food/royal jelly and enzymes for processing nectar into honey, 2nd the wax glands become operative at about 12d and finally the sting produces venom at about 18d.

It should be noted that the bee works for 8 hours, patrols for 8 hours and rests for 8 hours, although these activities are not performed in 8 hour stretches. The tasks undertaken by the worker bee at its various ages are generally referred to as the 'division of labour'.

3.12.2 Note rule of three!

3 castes queen, worker, drone.
Egg 3d to hatch.
Egg to worker emerging 3w.
Duty as house bee 3w.
Field bee 3w.
Life of drone c. 3m.
Life of queen c. 3y.

3.12.3 House cleaning.

This includes:

• Cell cleaning - removal of excreta, larval moults and then polishing cells ready for laying.
• Hive cleaning - removal of dead bees and debris from the floor. These are menial tasks undertaken by the youngest bees with no experience of other duties; they are performed instinctively and start more or less immediately the bee emerges from its cell.

3.12.4. Feeding/nursing.

• Feeding older larvae	3 - 6d.
• Feeding young larvae	6 - 9d.
• Capping brood cells	9 - 12d.

At 3d. old the hypopharyngeal glands start to become active and the nurse bees take up feeding duties for about 6d. (3-9d. old). At about 9d. old the wax making glands become active. Note very young and very old bees do not secrete wax. From about 9-12d old the nurse bees (now secreting wax) start brood capping duties. Note the colour of the cell cappings; they contain pollen mixed

with beeswax to make them porous allowing the larvae/pupae to breathe. Cells filled with honey are capped with pure wax.

3.12.5 Processing nectar into honey.

For a complete description see Section 3.13.

3.12.6 Wax making/comb building.

• Wax secretion generally occurs in bees 12-18 days old at relatively high temperatures 33^0-36ºC (91.4^0 - 96.8ºF). The wax is secreted in small flakes from 4 pairs of wax glands on the last 4 visible segments on the ventral side of the abdomen (colloquially known as the waistcoat pockets), ie. on the sternites A4 - A7.
• Large honey consumption is required in order that the bee may produce wax. In the literature various estimates are given; however 8lb of honey for 1lb of wax seems to be a realistic mean.
• When building comb, workers gorge themselves with honey and hang in festoons for c. 24 hrs. before the wax secretion and building process starts.
• A wax scale is removed by one hind leg and transferred to the mandibles by the two fore legs. The wax scale is thoroughly masticated before fixing to the comb and moulding it in place. When it is first deposited it is spongy and flaky and is later manipulated again making it smoother and more compact. Removing, masticating and fixing one scale takes about 4 minutes. It is not clear whether any secretions from any of the glands are used in comb building (mandibular, salivary, etc.). Some books indicate that the mandibular glands are used. On the basis of one scale taking 4 minutes to manipulate, 66,000 bee hours are involved building 77,000 worker cells using 1kg of beeswax.
• Bees can detect gravity (sensilla at the petiole between the head and the thorax) and the festooning chains (catenaries) play an important role in the parallelism of the combs.
• Queenlessness and bright light inhibit the bees building comb and secreting wax.
• According to Dadant, the thickness of the wall of newly built comb is approx. 0.0025in thick and in naturally drawn comb without the use of foundation, the base is 0.0035in thick; Hooper gives 0.006in for the cell wall thickness and Winston 0.073mm (0.003in).

3.12.7 Evaporation.

• The sugar content of nectar varies considerably depending on temperature, humidity. sunshine, wind speed and direction, etc. If the incoming nectar contains 40% sugar then there is c.60% water. After manipulation by the house bees, the ripened nectar will contain c.15% less water, ie. 45% water. Ripe honey contains c.20% water; the difference between 45% and 20% is due to evaporation as warm air is passed over the combs by fanning bees.
• Nectar is first manipulated by the bees (gorging and regurgitating), causing both a physical as well as a chemical change before it is hung up in droplets in empty or partially filled cells (ie. largest surface area). It is spread over a large area of comb and later gathered up and concentrated into a smaller area of comb. For this reason it is very important to provide adequate comb space during a nectar flow.
• On average it takes about 4/5 days to ripen nectar from 60% water content to honey of c.20% water content.

• During the journey back to the hive, the water content remains virtually unchanged.
• Evaporation is dependent on:

    - storage cells (space) available
    - temperature (directly)
    - humidity (inversely)
    - ventilation

• There is a need for a continuous stream of air from outside to inside as the air inside becomes saturated. The RH inside the hive varies from 20%-80%. In the brood nest it is fairly constant, 35%-45%RH.
• As an example during a heavy flow with a strong colony c. 2.5kg of water are evaporated in 24 hrs. or 50% of the gross gain per day. c.2/3 of this total loss occurring during the day.
• For each field bee (forager) there are approx. two bees in the hive for house duties which include the ripening and storage of honey. The importance of good hive ventilation is clear; it being important to allow top ventilation through the crown board and roof ventilators as well as the hive entrance at the bottom.

3.12.8 Ventilation.

• Ventilation of the hive is always necessary in order to expel CO_2, to expel water vapour when the colony is ripening honey and for cooling purposes in hot weather. This can be achieved by the bees but assistance can be given by the beekeeper so that the colony is put under minimum stress.
• When the temperature of the brood nest (93º-95ºF) tends to rise above its normal limits, there is a colony response. The bees sense the need for ventilation to control both the temperature and the humidity (35-45%RH). Temperature control is necessary to incubate the brood and control of RH is required so that brood food and royal jelly, when present, do not dry out.
• By evaporating water from nectar or from water actually brought into the hive, the heat is 'used up' thereby cooling the hive. When water evaporates there is a drop in temperature (known as the latent heat of vaporisation). The bees do this by fanning; it is a very economical and efficient air conditioning system.
• Fanning: bees collect to one side of the entrance and face inwards and fan vigorously creating an outward flow of air. Others are doing the same thing further in on the floorboard. There can be up to hundreds fanning and the draught created can easily be felt with the hand near the entrance. If the conditions become extreme, a further group will start fanning on the other side of the entrance but facing outwards and setting up in inward current of fresh cool air. Fanning is normally undertaken by older bees who are muscularly strong to perform this task.
• The beekeeper can assist in hot weather by:

    - providing sufficient space within the hive,
    - providing shade over the hive at noon and/or pm,
    - ensuring the crown board feed hole is open and the roof ventilators are not blocked,
    - keeping undergrowth trimmed around the hives,
    - in extreme conditions:- staggering supers, off-setting crown boards and roofs, raising hive above the floor (say 25mm or 1in).

-Note that late in the season to prevent robbing the hive must be bee-tight and only have one entrance which can be guarded.

• When bees are being moved the temperature increases very rapidly due to the disturbance and travelling screens are essential, particularly in summer.
• Ventilation in winter: The degree of ventilation required for successful wintering is not agreed among the experts and the literature can be confusing on the subject. The amount is determined by the sizes of the openings top and bottom and the convection currents (warm air rising around the cluster). The variables are:

     - size of entrance (nb. mouse guards),
     - size of top ventilation
        - raised crown board (matchsticks)
     - Morris board
     - crown board completely removed
     - size of roof ventilators.

It is generally agreed that greater ventilation is required in warmer and damper regions cf. colder and drier regions. However, it should be noted that about 4 gallons of water have to be dispensed with by evaporation due the consumption of c. 30lb of winter stores.

3.12.9 Temperature control.

• Individual bees have no means of controlling their body temperature and quickly assume the ambient air temperature (they are said to be poikilothermic). Their activity, both physical and physiological, quickens with a rise of temperature and slows as it falls. This automatic effect of temperature on metabolism may serve to stimulate equally automatic social actions to adjust the temperature of the hive. It should be noted that the bee is capable of sensing temperature, if it could not do so it would be unable to survive.
• Temperature is lowered by:

     - fanning at entrance and inside the hive,
     - water evaporation,
     - dispersion through the hive (as opposed to clustering),
     - clustering outside the hive entrance.

• Temperature is increased by:

     - muscular activity (thorax muscles),
     - clustering tightly,
     - manipulating a colony or moving a colony (muscular activity).

• Activity temperatures:

     - all activities occur between 10°-38°C (50° - 110°F) ,

- brood nest 35°C (95°F),
- unable to fly at 10°C (50°F) *,
- becomes immobile at 7°C (45°F),
- clustering starts at 14°C (57°F),
- thorax T= 20°-36° (68°-97°F) normally 29°C (84°F).

* Very often bees can be seen flying at air temperatures lower than this; the actual temperature of the bee is therefore above 10°C. Water collecting in the spring can be a very hazardous occupation for the honeybee; it must not allow its body temperature to fall below the critical 10°C while it is taking water or it can never return to the hive.

• The lower the temperature, the tighter the cluster (physically smaller) thereby providing a smaller surface area and less heat loss. At very low temperatures the bees bury their thoraces in the cluster and spread their wings (also to reduce heat loss) with abdomens out. The connective cluster under these conditions merges with the main cluster.
• The old adage that bees never freeze to death, only starve to death, is very accurate. With an adequate supply of stores they can survive very low temperatures by generating sufficient heat in the centre of the cluster to maintain the outer surface of the cluster at just above 7°C. This stops the bees on the outside from becoming immobile and falling off. Note that, contrary to popular belief, the bees in the cluster do not continually change position to keep warm.

3.12.10 Defence.

• Defence generally occurs at the hive or within a few metres of it. The defensive vigour of a colony depends on the genetic 'make up' of the strain, some bees being much more aggressive than others.
• No guard bees are likely to be found at the entrance of a colony during a nectar flow. Conversely, during times of dearth, many guard bees will be seen.
• Guard bees exhibit a typical stance - standing on their 4 rear legs, forelegs raised and antennae out-stretched. If some become alarmed, they open their mandibles and spread their wings ready for attack.
• Each guard bee 'patrols' a particular area. They check incoming foragers and challenge drifting bees by touching with their antennae (1-3 sec).
• Robbers (both bees and wasps) have a distinctive flight noticeable to guard bees who will always attack in defence rather than challenge first.
• Stinging is a defensive mechanism, not an attacking one. In an undisturbed colony less than $1/2$% of the bees in a colony are likely to sting (200 in 40,000).
• Guard bees are sensitive to:

- vibrations,
- visual stimulus (fast movements),
- odours (animals, humans and pheromones).

Once the guard bees have been alerted, they will fly round the area outside the hive, the distance they guard depending on the strain of bee. The Africanised bee in S. America will guard at distances greater than $1/2$ mile. Many bees in UK (usually bad tempered stocks) will follow for a few hundred yards. A few yards is typical for a reasonably tempered colony.

144

• Bees from the same colony have the same smell due to them having the same diet resulting from the food sharing and transmission among all the bees in the colony. For these reasons Bro. Adam believed there is a hive odour. Dr. Colin Butler on the other hand maintained there is a colony odour which is genetically produced. Whatever the reason it is generally agreed that guard bees can recognise bees from another hive whether they be drifters or robbers.

• Colonies working the same flower (eg. rape) would be expected to develop the same hive odour and recognition would be through colony odour. However under these conditions, where a flow exists, there are usually no guard bees present and it is clear more work is required on colony versus hive odour theories.

• Stimuli that elicit stinging behaviour are:

    - exhaled breath,
    - smell of hair, leather and many cosmetics,
    - violent vibrations and bumps,
    - rapid movements,
    - most important - alarm pheromones (bee venom and isopentyl acetate from the sting and 2-heptanone from the mandibular glands).

• Guard bees are usually 18d. old prior to becoming foragers. When the guards are alerted and pheromones are distributed around the hive, foraging tends to stop and the foragers become guard bees; many more bees are noticeable at the hive entrance under these conditions.

3.12.11 Foraging.

In a well balanced colony during the season c. $1/3$ worker bees are foragers, the other $2/3$ are house bees. Workers normally start foraging at c. 21 days old and continue until they die (away from the hive) c. 3 weeks later. They forage only in the daylight hours and in favourable weather; T55°F (13°C) and above, flying 6-10 ft. above ground in winds below 15mph (24 km/h). Flying at 15mph, they consume honey at a rate of 10mg/hr.

They forage up to c. 2.5 miles from the hive. They forage for nectar, pollen, propolis and water. When T=43°C (109°F) and above they only forage for water. A small number of the foragers (about 2%) act as scout bees, a very important activity; however it is virtually impossible to determine the exact number.

• Foraging is stimulated by:

    - presence of the queen and brood (both produce stimulating pheromones),
    - the needs of the colony indicated by food exchange in the colony and the speed foragers are unloaded by house bees,
    - scout bees dancing on the combs providing information on distance, direction and quality of the source. The most stimulating dances attract the most foraging activity.

• Foraging bees collect nectar, pollen, water and propolis.

3.12.12 Communication dances of worker honeybees.

• The main dances for communicating nectar and pollen sources, discovered by von Frisch, are:

- Round dance   -   sources up to 100 m
- Wagtail dance   -   sources over 100 m

• Round dance: contains little or no information except that the source is close to hand (within 100 m). Bees (newly recruited foragers) responding to this dance search in all directions from the hive. This dance is most apparent if wet supers are replaced on a colony after extraction during daylight. The colony very quickly (a few minutes) goes into a state of agitation with many bees 'milling around' the neighbourhood of the hive looking for the source which of course is on the hive above the brood chamber. This important 'deficiency' in the bees' communication system is the reason why wet supers should ONLY be returned to the hive after dark when the bees cannot fly.
• Wagtail dance: (see Appendix 14, page 339 in the Green Book) provides very precise information on the direction and distance from the hive to the source of forage and the time it is available.

- Direction is given as an angle between the food source and the sun. This angle is translated as the same angle between the vertical and the 'wagging direction' on the comb. The bee can determine the vertical by gravity sensors (between the head and thorax). The top of the comb always represents the sun. If the source, when viewed from the hive, is to the right of the sun, then the dance on the comb will be to the right of the vertical. Similarly for sources to the left of the sun, the dance will be to the left of the vertical.

- Distance information is given by the number of 'straight runs' (centre of the pattern) every 15 seconds as follows:

100 m = 9-10 runs/15 sec.
600 m = 7 ditto
1000 m = 4 ditto
6000 m = 2 ditto

- Wenner (1962) discovered that during the wagtail portion of the dance, a series of sound blips are made (250 Hz) which are inaudible to the human ear. The number of blips also correlate with distance. It is generally agreed that the number of runs per 15 sec is probably the most reliable indicator of distance.
- Time the forage is available is the time the bee is dancing on the comb. It should be noted that bees have the ability to 'remember' time (when the sun is in a particular direction).

• Other types of dances (which are poorly understood):

- Alarm dances (eg. poisoned food) - spirals or zig-zag.
- Cleaning dances - stamping legs + swinging body side to side.
- DVAV* ('joy') dances - front legs on another bee + 5/6 shaking movements.
- Massage dances - starts by bending head in curious way (sickness).
- Vibration dances - just before a swarm departs.
* = Dorsal - ventral - abdominal - vibration

3.12.13 The difference between summer and winter worker bees.

• In temperate regions where the winter dearth of forage is long, the temperate bee has developed to survive by clustering adjacent to a good reserve of food collected in the summer when forage is plentiful. In the winter cluster the metabolic rate of each bee is very low and little energy is being used. This is at variance with the tropical bee where the dearths of forage are much shorter and are not associated with the cold. These climatic differences have brought about a difference in life span of the different races and a difference in life span of seasonal bees in the temperate zones, ie. the summer and the winter bee.
• After emergence the life span of a worker bee can range from a few days to almost a year. This range of life span is dependent on the following:

a) seasonal factors, b) food availability, c) activities performed and d) the race of bee.

In the temperate zones the lives can be:
    Short lived (15 - 38 days) in the summer to long lived (c.140 days) in winter with
    Intermediate lives (30 - 60 days) again depending on a variety of factors

It should also be noted that bees in temperate zones respond more strongly to seasonal factors than tropical bees.

• Just after emergence the young bee consumes large amounts of pollen which causes the hypopharyngeal glands and the fat bodies to develop which provides the bee with a store of protein as brood food in the hypopharyngeal glands and in the fat bodies. Honey consumed is converted to glycogen and is also stored in the fat bodies. These protein stores in the hypopharyngeal glands and fat bodies can be used basically in two ways:

    a) In summer. The reserves are used in brood rearing by drawing down on the stored protein from the hypopharyngeal glands in the form of brood food. Maurizio (1950) showed that the more brood that a worker reared the shorter the life of the bee.
    b) In winter. The bees emerging in late autumn have very little brood to rear and the hypopharyngeal glands remain plump. The worker bees have many fat bodies as a result of the pollen consumption in the late autumn flows, eg. ivy in the UK. Most of the fat bodies in the winter bee are stored in the roof (dorsal side) of the abdomen. It has also been shown (by whom we know not) that the life of the bee is proportional to the amount of pollen consumed. This is evident if a colony suffers a dearth of pollen in the autumn then it is likely to succumb during the winter because of the shortage of fat bodies.

• From the above we can now say that the life of the bee (L) is proportional to the amount of pollen (P) consumed and inversely proportion to the amount of brood (B) reared. This is expressed more elegantly:

$$L \alpha P \div B \quad \text{or} \quad L = K_1(P/B) + K_2 \qquad \text{where } K_1 \text{ and } K_2 \text{ are constants.}$$

• In many ways the winter bee is similar to the bees in a queenless colony whereby the queenless workers revert to young bee types by consuming large amounts of pollen and re-developing the hypopharyngeal glands and fat bodies. However, there the difference ends because the queen, by providing queen substance prevents the enlargement of the workers' ovaries and therefore laying workers do not develop in the winter cluster.

• The winter bee has an important job to do in the spring to ensure the survival of the colony. It is its food reserves and its ability to live longer that makes this possible.

• It can be shown (Reference to diagram 4.4 page 310, the Green Book) that the variation of the percentage of bees in a colony with well developed hypopharyngeal glands and fat bodies during the course of the winter period is high and how in summer there is only about 20% with well developed glands.

• Taranov (1972) undertook some interesting studies on the lives of worker bees. We have not been able to discover where the work was done but it is stated to have been undertaken in a country with a rather warm winter and a hot summer so it may not be directly applicable to the UK. However, it is indicative of the general situation in our own hives.

EMERGE 3rd week	Average life span-summer	Average life span-winter	Average life span-spring	Average life span-total	%age BROOD cf. APRIL
April	3.5 weeks			3.5 weeks	100%
May	4			4	96
June	4			4	83
July	4.5			4.5	56
August	8	17 weeks	4 weeks	29	38
September	5.5	17	4.5	27	21
October	2	17	5	24	5

The results clearly show the extended life of the worker bees from August onwards being related directly to the amount of brood that they have to rear.

• The winter bee cannot depart the hive to take cleansing flights when the weather is inclement. The rectum can fill half the volume of the abdomen and extend throughout its whole length. The rectal pads are thought to be involved in the re-cycling of water from the stored faeces (Wigglesworth, 1932) and the build up of faeces in the rectum acts as an irritant which increases the temperature of the winter cluster. Noticeable drops in cluster temperature have been recorded after cleansing flights in winter.

• The nitrogen content of the winter bee's fat bodies is normally 14 to 23mg when healthy but decreases rapidly to about 6mg in bees infected with nosema.

• Finally, long cold winters are bad news for *Varroa destructor* because fewer infested bees survive until brood rearing starts again in the spring.

• In summary the difference between summer and winter bees are as follows:

Winter bee - long life - low fat content - high protein content - minimum brood reared - ability to retain high rectum content.

Summer bee - short life - high fat content - low protein content - maximum brood reared.

The longevity of queenless bees is due to the absence of brood rearing and the high level of protein in the hypopharyngeal glands and fat bodies.

3.13 The collection of nectar and how it is converted into honey suitable for storing in sealed comb.

3.13.1 General.

Nectar is collected from the nectaries and extra-floral nectaries of suitable flowers providing they are yielding nectar and the bee's proboscis is long enough to reach it.

- Nectar is sucked up by the proboscis, the average load being 40mg and the maximum load c. 80mg.
- The returning forager unloads by passing the nectar load to a house bee for manipulation and subsequent ripening.
- The speed of unloading by the house bee indicates the colony's needs (note that high sugar content is preferable to low sugar content).
- Foragers are constant to a particular species of flower and will remain in the hive if it is not yielding at a particular time.
- The number of trips per day is an average of 10 and generally range from c. 7-13 trips.
- The average time per trip is c.1 hour, half this time is spent flying and the other half collecting and unloading. When unfavourable conditions prevail, trips as long as 3-4 hours have been reported.
- The number of flowers visited per trip to obtain a full load is very variable (50-1000) and depends on all the variables associated with the secretion of nectar.
- It should be noted that bees also forage for honeydew, the sugary exudate of many species of aphid.

3.13.2 Constitution and types of nectar.

Nectar is the sweet secretion from the nectaries and extra-floral nectaries of flowers and plants. It consists of sugar (5 to 80%) and water plus very small quantities of the following substances that give it its characteristic colour and flavour when converted into honey:

proteins	mineral ash
amino acids	salts
enzymes	vitamins
lipids	mucus
organic acids	gums
vitamins	ethereal oils
alkaloids	dextrin
antioxidants	alcohols

149

Other external agents are always found in nectar such as pollen, fungi, yeasts and bacteria. It has a pH ranging from 2.7 to 6.4 (ie. acid to neutral) and is seldom alkaline.

Generally there are 4 types of nectar collected by the honeybee which are classified by the major constituent ie. the sugars:

Sucrose dominant (disaccharide $C_{12}H_{22}O_{11}$)
Glucose dominant (monosaccharide $C_6H_{12}O_6$)
Fructose dominant (monosaccharide $C_6H_{12}O_6$)
Balanced with equal amounts of sucrose, glucose and fructose.

Generally the composition of the nectar does not change with the age of the flower, between flowers or with variations in climatic conditions. It is also to be noted that glucose was formerly known as dextrose and fructose known as levulose. Fructose tastes twice as sweet as glucose and fructose can be tolerated by diabetics.

3.13.3 The conversion of nectar to honey.

The conversion involves two basic changes which are:

Chemical change - breaking down the sucrose with invertase, the predominant enzyme in nectar and which is also provided by the honeybee. The conversion process requires water which is present in the nectar as secreted:

$$C_{12}H_{22}O_{11} + H_2O \qquad + \text{ invertase} \qquad \rightarrow \qquad C_6H_{12}O_6 + C_6H_{12}O_6$$

sucrose + water + invertase $\rightarrow$ glucose + fructose

Physical change - evaporation of the excess water content to reduce it to about 18% for an average honey.

3.13.4 Conversion process by the honeybee.

• While the field bee is collecting nectar it adds invertase (= sucrase) from the hypopharyngeal glands. This secretion travels down the salivary canal of the proboscis where it mingles with the nectar and is then drawn up through the food canal and stored in the honey sac.
• The foraging field bee transfers its load to a house bee usually below and on the periphery of the brood nest.
• The house bee then moves its position away from the brood nest and manipulates* the nectar by regurgitating and swallowing again a large droplet of nectar. This operation takes about 5 to 10 seconds and during the process more invertase is added. It does this for about 20 minutes (say about 100 times). This manipulation process evaporates c. 15% of the water.
• The house bee then spreads the manipulated nectar to dry, on the upper surface of empty cells or into partially filled cells above the brood chamber.
*Note: The description by Gary and the drawings of the proboscis of the honeybee ripening honey by Janson under the direction of Park are, in our opinion, unsurpassed. They may be found in 'The Hive and the Honeybee' by Dadant.

• Fanning by other house bees removes the moist air from the hive to be replaced by dry air from outside. When the moisture content is down to c. 18% the cells are then sealed with wax cappings.
• Reference to Section 2.15.1.2 shows the amount of space required in the colony to convert nectar into honey.
• Other chemical processes are occurring during the conversion to honey and after it has been sealed. Higher order sugars are produced mainly from the glucose and the glucose oxidase reacts with the glucose to produce gluconic acid and hydrogen peroxide. The hydrogen peroxide acts against any bacteria that may be present and is the substance in honey responsible for its efficacy in the treatment of open wounds.

3.13.5 Honeydew honey.

The syllabus omits reference to honeydew which the honeybee can collect and store in the hive. However, we believe for completeness that mention should be made of the subject here; accordingly we have appended below from our manuscript notes some basic facts.

Insects, primarily of the order hemiptera (aphids) feed on the phloem sap from the phloem tubes of forest trees, notably *Pinus sylvestris* but also other conifers. The large quantities ingested by the aphids are quickly secreted as a sweet sticky substance called honeydew and if allowed to dry on the living parts of the plant is called 'manna'. Mixed in the honeydew and manna are the secretions from the salivary glands of the aphids which contain many more enzymes than nectar from floral sources and additional higher order sugars, examples shown below:

Enzymes - invertase, diastase, a peptidase, a proteinase, etc.
Sugars - melezitose, melibiose, erlose, raffinose, etc,

• The additional enzymes break down sugars such as sucrose, maltose and trehalose to produce sugars discrete to honeydew honey.
• Organic acids are always present (eg. citric and more rarely malic acid) and many amino acids (up to 22 have been identified in one sample of honeydew honey).
• In general, the honeybee converts honeydew and manna (usually collected early morning when it is diluted with dew) in the same way as floral nectars. The resulting honey can vary from a golden colour to nearly black; it would never be classed as light honey and the flavour is mild to very strong. The protein content is also higher. It fails to granulate evenly and is always marketed as a run honey. It is prone to deterioration due to the moulds on the honeydew before collection.
• The difference between honey from floral plant and honeydew and manna can be determined by the following formula:

$Q = -8.3x_1 - 12.3x_2 + 1.4x_3$ where
$x_1 = pH$, $x_2 = \%age$ ash content and $x_3 = \%age$ reducing sugars
If $Q < 73.1$ the sample is honeydew honey and
If $Q > 73.1$ the sample is honey from floral plants

It is well established that nectar from *Rhododendron ponticum* contains the toxin andromedotoxin acetylandromedol which is dangerous if consumed by man. Similarly there are honeydews which are also toxic to man and to quote one well known one from New Zealand ie. the secretions of the *Scolypopa australis* found on the *Coriaria arborea*.

In some countries honeydew honey is highly regarded and in others it is considered inferior. Most countries recognise it in their regulations regarding honey for sale.

3.13.6 Other points for consideration converting nectar to honey.

• It will be clear from the above that plenty of space is required and there is much sense in the old adage 'over super early in the season and under super late in the season'.
• It is important to provide conditions in the hive to allow the bees to ventilate and ripen their honey easily. The authors believe that by providing top ventilation it assists the bees to ventilate via the hole in the crown board and roof ventilators. In a nectar flow if the roof is raised there are always bees fanning around the open feed hole; we notice many beekeepers keep this hole closed for no apparent reason. It must be hard on the bees to move the air from the 3rd or 4th super down to the bottom entrance.

3.13.7 The interrelationship of nectar, honey and water in the honeybee colony.

The honeybee can only metabolise a sugar : water ratio of 50 : 50 to satisfy its own requirements and it strives to maintain this concentration continually throughout its life.

When nectar is being collected (sugar content varying from as low as 10% to as high as 60%) the house bee ripens this and then hangs it up to 'dry' by evaporation and fanning before sealing as honey with a sugar content of c. 80%.
If nectar is not coming into the hive then the colony must rely on its honey stores and these have to be diluted with water to 50 : 50 in order for them to be metabolised. Foragers start collecting water under these conditions.

The 50 : 50 ratio is required in order to metabolise sugar to provide the energy for feeding, heating, muscular energy for walking, synthesis of wax and the production of brood food and royal jelly.

3.14 The collection of pollen and how it is carried to the colony and used.

3.14.1 General.

Pollen is collected from the anthers of flowers either intentionally or by chance when foraging for nectar. It is collected by the plumose hairs covering the exoskeleton and transferred to the corbiculae by the 3 pairs of legs.

- The average load (weight of both pollen pellets) ranges from 11-29mg.
- The pollen collecting trips are completed more quickly than nectar collecting trips. Range is reported to be 3-18 minutes.

- The pollen forager returns to the hive and deposits the load directly into a cell adjacent to the brood nest without assistance from house bees. Later, house bees come along and pack the pollen loads into the cells and finally, for storing, they are sealed with a layer of honey and wax.
- Open brood provides a pheromone which induces foraging for pollen in addition to queen substance.
- The number of pollen foragers is controlled by the needs of the colony and the number of cells prepared by the house bees for pollen. It has been reported that they can vary from a few % to as high as 95% of the total foraging force.
- The number of foragers collecting mixed loads of both nectar and pollen is c. 3% of the foraging force.

3.14.2 The collection and use of pollen in the honeybee colony.

The colony needs a fertile queen and the pheromone from open brood to stimulate the foraging bees to collect pollen. Returning foragers recruit further foragers by dancing on the comb indicating to other bees the position of the source; the type of pollen is recognised by the aroma of the pollen on the bees' legs.

The number of pollen foragers can vary between wide limits depending on the colony requirements (eg. a few percent to as much as 90%).

When foraging, the bee alights on a flower and, moving quickly, bites the anthers of the stamen with her mandibles in order to dislodge the pollen grains. These pollen grains attach themselves to the plumose hairs which cover the whole of the exoskeleton. Then the bee leaves the flower and hovers nearby to clean the pollen from her body and to load it into her 'pollen baskets'. The process is as follows:

a) The front legs: by means of stiff hairs collect pollen from the head and first thoracic segment. The pollen is moistened by honey or nectar deposited on the front legs from the proboscis.
b) The middle legs: collect the pollen from the first legs and the rest of the thorax particularly the ventral side which is then passed on to the inner side of the basitarsi of the hind legs.
c) The hind legs: clean the abdomen and when sufficient pollen is collected on the inner surface of the basitarsi, these surfaces are raked by the 'pollen rake' at the bottom of the tibia of the other hind leg. The pollen is forced as a paste onto the flat surface of the auricle which is bevelled upwards and outwards. The tarsus closes against the tibia and the pollen is squeezed upwards and outwards onto the outside surface of the tibia. It is held in place here by the hairs on the corbicula, pollen basket, of the tibia (note the single hair acting as a pin through the load). One full load of two pellets represents approximately 100 flowers visited eg. dandelion when it is yielding well.

A few facts about pollen:

The average pollen load (both pellets) = 12 - 30mg.

The average trips per day = 6 - 8.
Total collected in one year = approximately 100 lb (45kg) per colony.
Amount required to raise one adult bee = 70 - 150mg.

• Storage of pollen. When the pollen forager returns to the hive with a load it has to be stored. She selects a cell near to the unsealed brood, grasps the edge of the cell with her forelegs and arches her abdomen so that the posterior end rests on the opposite side of the cell. The hind legs hang into the cell and the middle legs are used to push the pollen loads off the rear legs into the cell. The forager departs more or less immediately for another load. A house bee now comes along and breaks up the pollen and presses it firmly into the bottom of the cell with her mandibles. Honey or nectar is added to the pollen mass; it becomes darker, has a higher sugar content and is known as 'bee bread'.

The packed pollen can be fed to the brood or house bees (for producing brood food) or the cell can be filled with further loads topped off with honey and sealed with a wax capping for winter stores.

All pollen storage is adjacent to and around the brood nest where it is required for use though the odd cell or groups of cells filled with pollen are sometimes found in the supers.

It should be noted that after pollen has been collected by the bee, it is no longer viable for plant reproduction.

3.14.3 The use of pollen in the honeybee colony.
Pollen, the male germ cell of flowering plants (angiosperms), which comes from the anthers at the top of the filament of the stamens of flowers or catkins, has two major uses:

   a) it is the principal source of protein, fat and minerals in the honeybee diet,
   b) it can provide a surplus product from the apiary.

Pollen demand in the colony is related to the amount of unsealed brood. Bees cannot rear brood without pollen because the nurse bees cannot produce brood food from the hypopharyngeal glands unless the bees have consumed pollen. A strong colony will collect c. 50 - 100 lb (22 - 45kg) during a season.

   a) It requires 70 - 150mg of pollen to rear one adult bee.
   b) About 200,000 bees are reared during a season thus accounting for more than 50% of the income.
   c) The balance is used by the adult bees preparing for winter (increasing their fat bodies) and/ or stored in the comb for use early the following year before new supplies become available.
   d) Note the weight of a worker bee = 90mg therefore 1 pound of bees contains c. 5000 bees.

Pollen is rich in protein and is essential for body building material for growth/development and for the repair of worn out tissue. It also has the very important function of stimulating the development of the hypopharyngeal glands and the fat bodies of the winter bee. The protein content varies between different pollen types and also from flower to flower in the same foraging area. A protein content of c. 35% is typical of a high protein pollen eg. beans. Bees can discriminate between pollens by colour and odour; they cannot distinguish between the quality (protein content) of various pollens. There are wide variations in the protein content of different pollens and the bee more than likely receives a balanced diet due to the variety of pollens collected and used.

The use of pollen for brood rearing:

- Worker larvae are fed with brood food only from 0 to 3d. and then on 4th and 5th day with pollen, honey and brood food.
- Queens, both adult and larvae, are fed exclusively on royal jelly.
- After emergence of the worker bee, pollen is essential for it to reach maturity in a healthy state. It depends on pollen for the orderly development of its glandular system while it is a house bee.

In areas where natural pollen is in short supply, particularly in the spring, pollen patties can be fed to colonies to stimulate spring build up. Pollen shortage often occurs where colonies are foraging on honeydew in pine woods.

3.15 The production of wax and how it is used in the colony.

3.15.1 Wax glands and wax production. See section 3.3.4

3.15.2 The initiation of comb building and the construction of comb.

3.15.2.1 Initiation of comb building.

If a swarm is to survive when it arrives at a new cavity where there is no old comb, it must rear new worker bees and collect sufficient stores for winter. In order to achieve this, comb must be built to provide storage for honey, pollen and young brood. Seeley and Morse have calculated that a completed nest has comb with 100,000 cells with a comb surface area of $25,000cm^2$ (ie. approx 20 B.S. frames) and this needs 1200g of beeswax (ie. 2.6 lbs).
It has been shown by Combs (1972) that the average bee in a swarm carries 35mg of honey with a 65% sugar content and thus a swarm of 12,000 bees carry 273g sugar.

> 1g beeswax = $20cm^2$ comb
> 100,000 cells = 1,200g beeswax = 7,500g honey = 6,000g sugar.
> 1g sugar converts to 0.2g beeswax, therefore 273g sugar = 54.6g beeswax

Hence the swarm carries reserves to build $1,100cm^2$ of comb or c.5 % of the total colony requirements (54.6 ÷ 1,200 = 4.55% or 273 ÷ 6,000 = 4.55%).
This reserve is very meagre and represents the ability of the swarm to build about ½ BS frame from the resources it carries with it. Therefore, it is essential that the swarm starts to forage successfully more or less straight away and not be restricted by inclement weather.
- Comb building is inhibited by bright light and queenlessness. A queen is essential and the light

explains partially (temperature is the other reason) why comb building is always initiated in the middle of the cluster.

• The swarm starts by removing all loose particles from the top part of the cavity and then cleaning it.

• The bees cling to the top and hang in festoons metabolising the stored sugar brought with them. 24 hours later wax scales are produced and comb building commences.

3.15.2.2 Comb building.

• The wax scales are removed from the wax pockets by the middle leg and then passed to the mandibles with the two front legs.

• The scale is chewed in the mandibles together with secretions from the mandibular glands. It is not known what part the secretions play in the wax building process.

• The scales are then deposited on the upper inside surface of the cavity to form a small ridge of wax 2 to 4 mm in height.

• Hexagonal depressions are then moulded into the sides of the ridge and the excess wax is deposited along the sides of the hole to form the walls of the cells. The same process is carried out on the other side to form the double sided comb.

• The comb, as it is built downwards, has the last few cells of ever decreasing depth in the form of a taper down to the septum, the last few millimetres of which are in the formation of the initial thick ridge of wax.

• The antennae are instrumental in achieving the very precise cell size which is uniformly built; bees with amputated antennae cannot build uniform comb true to normal size.

• In the natural state the shape of the comb built is that due to the shape of the bees hanging in festoons which mathematically must be that of the catenary curve; ie. the natural curve of a chain suspended from its two ends, the distance apart of these two ends and the tension in the chain governs the shape of the catenary curve. The shape depends on the tension in the chain, the greater the tension the flatter the curve.

• The combs are built vertically and this is achieved by the use of gravity. The bee has special sensilla and associated setae between the head and the thorax and additionally at the petiole between the thorax and the abdomen. When the bee is in a vertical position with its head at the top any deviation from the vertical will throw its head either forwards or backwards which is sensed by the special sensilla. Similarly the process is identical at the petiole.

• The bees have a preference for the orientation of the combs in the horizontal plane. It has been found that they will build naturally at an angle to the magnetic meridian of about $50°$ in a NE - SW direction. The angle has been found to vary between different species of bees.

• The number of worker cells = c. 5 per inch (25mm) and those for drone = c. 4 per inch (25mm).

• The depth of the worker brood cells is quite constant and = 0.437in (11mm) giving the thickness of the comb as 0.875in (22mm). The depth of cells used for honey storage can increase to as much as 0.625in (16mm).

• The spacing between combs, which are always built parallel to one another, is 1.375in (35mm) where brood is being reared which gives a spacing between adjacent comb faces of 0.5in (12mm).

• The thickness of the wax walls is variously quoted between 0.0025in (0.06mm) and 0.007in (0.18mm) with the thickness of the septum not varying as widely and being c. 0.008in (0.2mm).

• It should be noted that left to its own devices the honeybee will always build its comb to suit the shape and dimensions of the cavity adopted as its nesting site.

3.15.3 Other uses of wax in the colony.

a) Capping cells filled with ripened honey and for capping pollen (pickled) for winter storage. The cappings for both these purposes is with pure beeswax to exclude any moisture.
b) Capping cells containing brood. These cappings are a mixture of pollen and beeswax to make them porous in order that larval respiration can take place and to make them attractive to the adult bee to chew its way out at emergence.
c) For building queen cells.
d) For use as brace and burr comb to give rigidity to comb forming the nest site.

3.16 The collection of water and propolis and how they used in the hive.

3.16 1 The collection and use of water by the honeybee colony.

This is the only item that is collected by the honeybee which is not stored in the hive. It is collected in the same way as nectar with the proboscis and transported to the hive in the crop.

- Water is required for diluting honey so that it can be metabolised, diluting honey for brood food for larval feeding, for humidifying the brood nest and for cooling the hive in very hot weather (latent heat of vaporisation) which is deposited on top bars and crevices around the hive.
- The average load is 25-50mg. and the foraging trip is very short, most are completed in under 10 minutes.
- The average colony in the spring requires c.150g. per day and a strong colony in hot drought conditions requires c. 1kg per day.
- Reception by the house bees indicates the colony needs; if unloading takes longer than 3 minutes then water collection activity ceases.
- Water foragers mark favourable sites with Nasonov gland and fanning.
- At times water is stored, not directly in the hive but in the crops of the receiving house bees (reservoir bees); this happens when supplies are not readily available.

a) The annual consumption is c. 44lb (4.4 gallons).
b) Large amounts required for brood rearing (c. 200 g/day)
c) Only small amounts required when over-wintering.
d) Water foraging trips are short (c.50 to 100/day with loads of c. 25mg)

3.16.2 The collection and use of propolis by the honeybee colony.

3.16.2.1 Definition of propolis.

• Propolis is a resinous substance obtained by the honeybee from trees and from the exudation of wounds in many types of woody plants. Examples of trees where propolis is regularly found are horse chestnut (the sticky buds), alder, poplar and the cherry family.

• Propolis contains an indeterminate number of substances and therefore has no discrete chemical formula. Some of the substances have antibacterial qualities which have a direct effect on some of the uses of propolis by both the honeybees and by man.

• The colour ranges from orange to dark brown or nearly black.

• It has a characteristic smell which is difficult to describe (clean and antiseptic-like?).

• It is waterproof and does not dissolve in water. It does dissolve in methylated spirit, alcohol and ether.

• We have not been able to trace the melting point but at summer temperatures within the hive it is extremely sticky.

3.16.2.2 The collection of propolis.

• The amount of propolis collected by a colony varies widely depending on the race and strain of bee.

• Because it is hard and brittle at low temperatures, propolis is only collected in warm weather thereby making the task much easier.

• The propolis foragers are 'old' bees. It is collected as follows:

    - biting and pulling off a small piece,
    - kneading it in the mandibles,
    - transfer from mandibles to pollen basket with middle leg on same side,
    - repeats on the other side,
    - pats the loads in to position with the middle leg.

• On return to the hive the forager is unloaded by a house bee which takes about 1 hour and involves biting and pulling off small pieces until the unloading is complete.

• The unloading process is prolonged because the propolis load is used at the point of requirement straight away. It is not stored for future use.

• The forager is faithful to her work and returns for another load.

• We have been unable to trace any information on how long it takes the forager to collect a load of propolis, how many trips per day they make under normal circumstances and an accurate figure for the weight of an average load.

• In the absence of available sources of propolis, honeybees will collect man-made alternatives such as tar from roadwork, paints and varnishes, caulking compounds, etc. We had tar collected by all the nuclei in our mating apiary one year for some reason best known to the bees; curious because the apiary has been established for some years and it has never happened again. Also we know of one case where the bees were collecting caulking compound from the next door neighbour's double glazing installation.

3.16.2.3 The uses of propolis.

• Inside the hive:

    a) For varnishing the inside surfaces of brood cells before polishing for the queen to lay in them. The antibacterial qualities of propolis preserve the brood food and minimise the spread of infections where brood is being incubated en masse.

b) All rough surfaces are propolised presumably to provide a smooth surface. The entrance to the hive also receives this treatment. The reason is unknown.

c) For filling cracks, to keep out draughts and rain.

d) The antibacterial qualities prevent, to some extent, the spread of wood rot.

e) Embalming dead intruders such as mice, small snakes, etc. The antibacterial effects prevent decay while the corpse is dehydrating.

f) Reducing the size of the entrance (nb. pro = before and polis = city). On some occasions we have had colonies that have built a wall of propolis right across the entrance using considerable quantities, a solid block about 1in × ½in × 16in (25mm × 12mm × 406mm).

g) When 'bee space' is not observed by the beekeeper any gap less than 6mm (¼ inch) will be filled with propolis.

h) When hives are moved, propolis reduces movement of the frames and other hive parts.

• By man:

a) As a varnish particularly by violin makers.

b) For dressing wounds; the propolis is cleaned, frozen and then ground into a powder and mixed as a salve with a suitable base substance.

c) Similarly it can be dissolved in spirit to make up a tincture. Recent experiments at a London teaching hospital reported in 'The Lancet' that a high success rate had been achieved treating internal ulcers when 2 or 3 drops have been added to a tumbler of milk taken regularly each evening for 3 months.

d) It can also be made into capsules to take internally.

e) Many beekeepers advocate sucking a small piece to cure a sore throat.

f) It is extensively used in Russia for veterinary treatment of livestock disorders.

3.16.2.4 Other points about propolis.

• Some cases have been reported of beekeepers developing an allergy to propolis. This is overcome by wearing surgical gloves for manipulating the colony.

• There is no doubt that beekeepers handling many colonies dislike those colonies which collect propolis in abundance. It makes the manipulation more difficult and more time consuming. It is possible to breed for minimum propolis; the best colonies that we have seen were those maintained by Brother Adam when he was responsible for bee breeding at the Abbey. His colonies appeared to collect just sufficient to treat the brood cells.

3.17 The factors that may give rise to swarm, supersedure and emergency queen cells.

3.17.1 Queen substance.

It is now known that the pheromone queen substance, a complex mixture (31 found) of mainly fatty acids of which 13 have been identified, is solely responsible for the building of all types of queen cells. It is produced by the mandibular glands of the queen and also from other dermal glands on her abdomen. It has not been possible to synthesise it successfully to prevent swarming although it was attempted by Glaxo some years ago. The following points in relation to the pheromone are important when considering the swarming behaviour of the colony:

• Butler showed that the behaviour pattern of the colony depended on the actual amount the colony received, ie. it is a quantitative problem.
• J.B.Free has stated that a young (new) queen produces about $5000\mu g$/day.
• A minimum threshold amount is required by each worker bee to prevent the building of queen cells. It is unclear whether the threshold amount is the same for all races and strains of bee. Similarly, it is unclear whether different queens produce the same amount per day. Compare the large prolific yellow colony with the smaller less prolific black colony. Does the yellow queen produce more QS or do the yellow workers require a lower threshold? Conversely, does the black colony require a higher threshold or does the black queen produce less?
• QS produced by the queen decreases with time and obeys the exponential law of decay. If she produces an average of $5000\mu g$/day during her first year then this will halve ($2500\mu g$/day) during the second year and halve again ($1250\mu g$/day) in the third year, etc. The importance of maintaining a young queen to prevent swarming will be obvious.

3.17.2 The start of swarming in a colony.

When the supply of queen substance is below the threshold required for colony cohesion, the queen's egg laying rate will rapidly decrease because of reduced feeding of the queen by the workers. Those eggs which have been laid in the queen cups, which are a part of every normal colony, will not be removed but will be allowed to hatch out into larvae. Queen cells will result and the colony will be on its way to swarming. Other signs will be apparent as follows:

a) House bees will be reluctant to accept nectar loads from foragers.
b) Foraging diminishes and redundant foragers start to seek a new nesting site.
c) The queen ceases to be fed and decreases in weight by c. 30% to enable her to fly.
d) Egg laying virtually stops.
e) The decrease in foraging and brood rearing results in physiological changes to the worker bees. Because of the reduced level of QS, worker ovaries start to develop and also because of lack of brood rearing the hypopharyngeal glands also develop producing additional fat bodies and an additional protein reserve.

The number of queen cells appears to depend very much on the strain of bee; those strains prone to swarming and genetically inclined that way will build very large numbers. We have counted as many as 70 in such colonies. For the more normal colony it is likely to be between 10 and 20.

By the time the first queen cell is sealed there will be no freshly laid eggs in the colony and the queen will be physically able to fly. Assuming the weather to be favourable the swarm will emerge at about noon. The following events are pertinent:

a) The emergence is preceded by the 'whir' or 'buzz' dance where the bees run backwards and forwards across the combs in horizontal lines buzzing with half spread wings every 0.5 to 3.0 seconds. Buzz frequency is c. 180 to 250 cps.

b) The dancers touch other bees (for up to 5 seconds) buzzing at 400 to 500 cps when touching.

c) The disturbance and excitement multiply and soon lead to the emergence of the swarm. Exactly which bees go is unknown and some return very quickly to the hive.

d) We have not been able to trace at what stage the bees gorge themselves with honey before departure from the hive.

e) The prime swarm with the old queen will settle usually within 10 to 20 metres of the hive.

f) Queen pheromone and Nasonov pheromone are vital to co-ordinate the swarm while in flight and during the settling process. It should be noted that the supply of queen pheromone from the queen though inadequate in the parent colony will be adequate in the swarm because of the reduction in the number of workers.

3.17.3 Factors giving rise to supersedure cells.

3.17.3.1 Definition of supersedure.

The definition of supersedure is the requeening of the honeybee colony without the colony swarming. There are two types of supersedure as follows:

> **1. Perfect supersedure** - whereby the new queen becomes mated and starts laying in the colony together with the old queen before the bees dispose of the old queen. Occasionally the two queens are seen. We have only observed it once or twice.
> **2. Imperfect supersedure** - whereby the bees dispose of the old queen before the new queen is laying.

3.17.3.2 Signs of supersedure.

• The signs within the colony are often very similar to swarming but the number of queen cells tends to be less than with swarming. The same queen cups, built during normal colony development, are used.

• Note the difference in the position of swarm and supersedure cells which are identical when compared with emergency queen cells which are built on worker larvae amid worker brood. Some books indicate that the queen cells are at the top of the brood nest for swarming and at the bottom for supersedure; we do not consider this to be a useful guide.

• Similarly, the number of queen cells has been quoted as being a guide with 1 to 10 for supersedure and 5 to 20 for swarming. Again we believe such figures can be unreliable.

• It is impossible to tell whether the bees will swarm or supersede or indeed destroy the cells altogether and do neither. Beekeepers must always assume that the colony intent is to swarm.

• The number of colonies that supersede is, in our opinion, about 10%. We base this on the number of new queens that we find at our first spring inspection; ie. unmarked and without clipped wings. The beekeepers who never mark and clip their queens will be unaware of this state of affairs.

3.17.3.3 Reasons for supersedure.

Many reasons have been suggested for supersedure; some are as follows:

a) Presence of Nosema disease (Farrar 1947). There is no doubt that queens are affected by the disease. We have dissected queens and checked for positive signs of Nosema and then found the spermatheca to be abnormal with a dark black colour instead of the shiny silvery colour of the surrounding glistening tracheoles. We have never ascertained what this condition meant.

b) An inadequate supply of eggs (Wedmore 1942 and Root 1948). It is not clear how these two came to this conclusion. However, it does suggest that the queen is inadequate for whatever reason.

c) Physical damage to the queen (Wedmore 1942 and Snelgrove 1946). In our opinion highly likely but no proof can be offered to substantiate the statements.

d) High drone population and shortage of sperm (Snelgrove 1946). Again no proof offered; in our opinion shortage of sperm usually manifests itself in a drone-laying queen rather than a supersedure, all other things being equal.

e) Shortage of queen substance (Butler 1954 - 56). Which seems to be the most realistic of them all as it will lead to the production of queen cells but how the bees deal with them is still unknown ie. to swarm or to supersede.

3.17.3.4 Other points on supersedure.

• It would be interesting to know the incidence of supersedure in the African bee which is a more prolific swarmer compared with its European counterpart.
• In the ultimate it would seem that all colonies must supersede their queens at some time or another. When a colony swarms with an old queen due to reduced QS, initially when it sets up a new nest the amount of QS will be adequate for the lesser number of bees in the swarm cf. the old parent colony. Once the new colony is established to a size whereby it can survive, in the ultimate a supersedure must occur. Little or no work has been done to our knowledge on this very important point.

3.17.4 Factors giving rise to emergency queen cells.

3.17.4.1 Description of a queen cell.

Unlike worker and drone cells, a queen cell hangs vertically no matter what sort it may be, swarm, supersedure or emergency. It is found sometimes between the combs in the bee-space but more usually on the comb or towards or on its edges. It has a pitted appearance (rather like the dried pod of peanut) on its exterior, tapering in shape towards its tip about 1 inch (25mm) in length and about ½ inch (12mm) in diameter at its upper end. It contains a cocoon which fills about the lower half of the cell and this cocoon does not make contact with the tip of the cell where the queen emerges by chewing around the periphery of the cell tip. The colour is best described as coffee, virtually identical with the colour of healthy brood cappings.

3.17.4.2 Queen cups.

Before dealing with emergency queen cells it is well to consider queen cups. In some parts of the country they are known as 'play cells'; we have never managed to fathom out the origin of this name. All swarm and supersedure cells start life as a queen cup. They are built by all *Apis mellifera* colonies in the spring and those which are not used are destroyed at the end of the season and new ones are rebuilt the following spring.

We have seen, and still do see, many beekeepers destroying these queen cups during normal colony inspections in the belief that they are preventing the colony swarming. We are amazed at these beekeepers and often wonder what other things they get up to in their ignorance of colony behaviour. Often an egg is to be found in these queen cups; this is not necessarily a sign of swarming or of supersedure, it is often removed by the worker bees. Only when a very young larva can be seen, floating in a pool of royal jelly in the queen cup, can one be sure that the swarming or supersedure impulse is starting to manifest itself.

3.17.4.3 Emergency queen cells.

Emergency queen cells are only built by a colony when it has lost its queen either by accident or design of the beekeeper. It is unusual for a feral colony to find itself in such a state. Once the queen has been removed from a colony it takes about 5 or 6 minutes for the colony to recognise that she is missing and the first signs of emergency queen cells being built will be visible in about one or two hours. To produce a viable queen it is necessary for the bees to build on cells containing young larvae; a larva less than 36 hours old will produce a good queen whereas a larva greater than 36 hours old will produce a poor queen. The younger the larva the better and the older the worse.

Suitable larvae are found in the brood nest. After the emergency queen cells are sealed, they are usually surrounded by sealed worker brood. The laying pattern of the queen is concentric in an ever increasing diameter, thus the youngest larvae will always be on the outer edge of a given laying pattern. Emergency queen cells are never likely to be found away from the brood nest or hanging on the bottom of frames and comb. Emergency queen cells are easily missed unless the bees have been removed from the comb.

A worker cell is extended outwards and downwards and fed with great quantities of royal jelly, so much in fact to enable the larva to be floated out to the base of the vertical section of the emergency queen cell. Sometimes, it is stated, emergency queen cells are built on drone cells with drone larvae and very often the drone larvae fall to the mouth of the cell. Such cells are characterised by their excessive length. Such drone larvae do not survive past the larval stage.

It should be noted that harsh swarm prevention methods (eg. Demareeing) are likely to induce the building of emergency queen cells. The mechanism is obvious whereby part of the colony (at the top) is separated from the queen and brood chamber below by one or more supers of honey, thereby creating a shortage of queen substance in the upper chamber. A similar mechanism may be experienced while undertaking a Bailey frame change (See Appendix 9, Beekeeping Study Notes Volume 1 - the Green Book).

3.17.4.4 Other points of interest.

The habit of bees to reseal the queen cell from which a virgin queen has recently emerged should be noted.

A ripe queen cell (within say 36 hours of emergence) is characterised by the darkening of the tip.

3.18 The use made by honeybees of the alarm pheromones and the effect these have on the way bees are managed.

3.18.1 General.

We are of the opinion that the effect of the alarm pheromones have no effect on the way bees are 'managed' by the beekeeper, as will be shown below in the following sub-sections, but they do have an effect on the way bees are 'manipulated' by the beekeeper. The only exception is with honeybees that demonstrate a strong defensive trait whereby they could well induce the beekeeper to become a leave alone beekeeper and not manage them at all! There are two alarm pheromones used almost exclusively for defence of the colony, their release being instinctively occasioned by the honeybee; the honeybee does not make conscious use of alarm pheromones.

There are two alarm pheromones namely:

> **2-heptanone**, which is derived from the mandibular glands of the worker.
> **isopentyl acetate**, which is the major component of many others (24 have been identified), is very powerful and derived from glands in the sting chamber.

The alarm pheromones inducing the honeybee to sting requires a threshold amount to elicit this effect. In docile bees the threshold level is high and in aggressive bees the threshold level is low. In the latter case this can lead to mass attacks seldom occasioned in the UK but more often in other countries with Africanised bees. With mass attacks where 500 or more stings are recorded death can occur even though the patient is not hypersensitive, however, the chances of developing hypersensitivity, after recovery, are increased. One of the Authors suffered a mass attack many years ago but did not become hypersensitive but it did take 4 or 5 days in bed to recover.

3.18.2 Defence of the colony.

• Defence generally occurs at the hive or within a few metres of it. The defensive vigour of a colony depends on the genetic 'make up' of the strain, some bees being much more aggressive than others.
• No guard bees are likely to be found at the entrance of a colony during a nectar flow. Conversely, during times of dearth, many guard bees will be seen.
• Guard bees exhibit a typical stance - standing on their 4 rear legs, forelegs raised and antennae outstretched. If some become alarmed, they open their mandibles and spread their wings ready for attack.
• Each guard bee 'patrols' a particular area. They check incoming foragers and challenge drifting bees by touching with their antennae (1-3 sec).
• Robbers (both bees and wasps) have a distinctive flight noticeable to guard bees who will always

attack in defence rather than challenge first.

• Stinging is a defensive mechanism, not an attacking one. In an undisturbed colony less than $1/2$% of the bees in a colony are likely to sting (200 in 40,000).

• Guard bees are sensitive to:

    - vibrations,
    - visual stimulus (fast movements),
    - odours (animals, humans and pheromones).

Once the guard bees have been alerted, they will fly round the area outside the hive, the distance they guard depending on the strain of bee. The Africanised bee in S. America will guard at distances greater than $1/2$ mile. Many bees in UK (usually bad tempered stocks) will follow for a few hundred yards. A few yards is typical for a reasonably tempered colony.

• Bees from the same colony have the same smell due to them having the same diet resulting from the food sharing and transmission among all the bees in the colony. For these reasons Bro. Adam believed there is a hive odour. Dr. Colin Butler on the other hand maintained there is a colony odour which is genetically produced. Whatever the reason it is generally agreed that guard bees can recognise bees from another hive whether they be drifters or robbers.

• Colonies working the same flower (eg. rape) would be expected to develop the same hive odour and recognition would be through colony odour. However under these conditions, where a flow exists, there are usually no guard bees present and it is clear more work is required on colony versus hive odour theories.

• Stimuli that elicit stinging behaviour are:

    - exhaled breath,
    - smell of hair, leather and many cosmetics,
    - violent vibrations and bumps,
    - rapid movements,
    - most important - alarm pheromones (bee venom and isopentyl acetate from the sting chamber and 2-heptanone from the mandibular glands).

• Guard bees are usually 18d. old prior to becoming foragers. When the guards are alerted and pheromones are distributed around the hive, foraging tends to stop and the foragers become guard bees; many more bees are noticeable at the hive entrance under these conditions.

3.18.3 The alarm pheromones.

3.18.3.1 The definition of a pheromone.

A pheromone is a chemical substance, secreted from the exocrine gland of an animal, that elicits a behavioural or physiological response by another animal of the same species and so acts as a chemical message. It is secreted as a liquid and transmitted as a liquid or a gas.

3.18.3.2 Alarm pheromone (Isopentyl acetate). Is derived from glands in the sting chamber of the worker honeybee.

• The glands are believed to be located on the inner surfaces of the sting quadrate plates and the pheromone from the sting chamber collects beneath the sting sheath on the bristle covered membrane which folds over the bulb of the sting shaft. The secretion appears to originate from two masses of glandular cells on the inner surfaces of the quadrate plates (J.B.Free).

• Isopentylacetate (a very powerful alarm pheromone) is in evidence whenever the sting is used. The pheromone has three effects, firstly to elicit a stinging response from other bees, secondly to mark the area to be attacked and thirdly to reduce the foraging force which reverts to defence duties. The quantity is small (1 to 5µg in workers c.15 to 30 days old). The glands that produce the pheromone have not yet been identified but it is known that the secretion collects on the bristly membrane at the base of the stylets near to the bulb.

• The pheromone is a powerful alarm pheromone the main constituent being isopentyl acetate which inhibits foraging and scenting. Many other components (24) have been identified but their functions have not been fully determined. The amount involved is as follows:

Young bees to 3 days old	nil,
2 to 3 weeks old	increases to a maximum = 4 to 5µg,
3 to 6 weeks old	decreases to about 2µg per sting.

• One important constituent is (Z)-11-Eicosen-1-ol which strongly inhibits scenting (helps to retain the more volatile constituents such as isopentyl acetate) and significantly attracts foragers. This goes a long way in explaining the old adage that 'the first sting is the most expensive one'. (Z)-11-Eicosen-1-ol attracts the alerted guards to the site of the first sting and the isopentyl acetate elicits the stinging around the same spot.

• It is used in two ways:

a) By opening the sting chamber with the abdomen orientated upward and the sting exposed. This is usually combined with fanning to disperse the pheromone and is noticeable if a colony is opened in cold weather and the cluster is formed.

b) When the bee has stung, the sting and associated motor apparatus remains in the victim.

• This alarm pheromone elicits a stinging response in other bees and recruits other workers to act as guard bees. It is to be noted that bee venom does not elicit a stinging response, only the alarm pheromones do this, ie. isopentyl acetate and 2-heptanone.

3.18.3.3 Alarm pheromone (2-heptanone).

Is derived from the mandibular glands of the worker honeybee.
The function of the glands changes with the age of the bee and the duties it is undertaking. 2-heptanone is not found in the secretion of very young bees. Only when the bees are of foraging age does the pheromone appear, ie. after they have finished their nurse bee duties. The total amount in an adult bee is estimated as up to 40µg and two other uses have been postulated as follows:

1. Marking exhausted forage.
2. Indicating larvae that have just been fed although this seems to be at variance with it being found only in older bees.

3.18.4 The effect of alarm pheromones on colony manipulation.

Once the colony has become alarmed and starts attacking the manipulator by stinging, isopentyl acetate is released in ever increasing quantities and those with an acute sense of smell will recognise the characteristic aroma of banana oil. At such a stage the colony becomes out of control and should be closed down as quickly as possible.

When manipulating a colony of honeybees, due cognizance should be taken of the points listed above in Section 3.18.2 to avoid alarming the colony.

** ** ** **

FORAGING

The Candidate will be able to demonstrate understanding of:

4.1 The main plants of local importance to the bees throughout the year, giving details of flowering times.

The main flora will vary from district to district throughout the UK and for examination purposes the Examiner will only concentrate on the local flora. Below is a general list based on the flowering times; select the ones which are applicable to your area and your apiary.

COMMON NAME	PROPER NAME	FAMILY	NECTAR or POLLEN
February/March			
Snowdrop	*Galanthus nivalis*	*Amaryllidaceae*	P
Crocus	*Crocus spp.*	*Iridaceae*	P
Gorse	*Ulex europaeus*	*Leguminosae*	P
Hazel	*Corylus avellana*	*Corylaceae*	P
Willow(goat)	*Salix caprea*	*Salicaceae*	P
Yew	*Taxus baccata*	*Taxaceae*	P
March/April/May			
Blackthorn(sloe)	*Prunus spinosa*	*Rosaceae*	N+P
Dandelion	*Taraxacum spp* *	*Asteraceae*	N+P
Gooseberry	*Ribes uva-crispa*	*Grossulariaceae*	N
Currants	*Ribes spp*	*Grossulariaceae*	N
Rape	*Brassica napus*	*Cruciferae*	N+P
Top fruit	**	*Rosaceae*	N+P
Bluebell	*Endymion non-scriptus*	*Liliaceae*	N+P
Sycamore	*Acer pseudoplatanus*	*Aceraceae*	N+P
H. Chestnut (Wh.)	*Aesculus hippocastanum*	*Hippocastanaceae*	N+P
H. Chestnut (Red)	*Aesculus carnea*	*Hippocastanaceae*	N+P
Hawthorn	*Crataegus monogyna*	*Rosaceae*	N+P
Holly	*Ilex aquifolium*	*Aquifoliaceae*	N+P
Mountain ash	*Sorbus aucuparia*	*Rosaceae*	N+P
Laurel	*Prunus laurocerasus*	*Rosaceae*	N+P
June/July/August			
Poppy	*Papaver rhoeas*	*Papaveraceae*	P
Thistle	*Cirsium arvense*	*Asteraceae*	N+P
Hogweed	*Heracleum sphondylium*	*Umbelliferae*	N+P
Field bean	*Vicia faba*	*Leguminosae*	N+P
Raspberry	*Rubus idaeus*	*Rosaceae*	N+P
White clover	*Trifolium repens*	*Leguminosae*	N+P
Charlock	*Sinapis arvensis*	*Cruciferae*	N+P

COMMON NAME	PROPER NAME	FAMILY	NECTAR or POLLEN
Runner bean	*Phaseolus multiflorus*	*Leguminosae*	N+P
Lime	*Tilia vulgaris*	*Tiliaceae*	N+P
Blackberry	*Rubus fruticosus*	*Rosaceae*	N+P
Willow herb	*Epilobium angustifolium*	*Onagraceae*	N+P
Bell heather	*Erica cinerea*	*Ericaceae*	N+P

August/September

Evening primrose	*Oenothera biennis*	*Onagraceae*	P
Ling	*Calluna vulgaris*	*Ericaceae*	N+P
Old man's beard	*Clematis vitalba*	*Ranunculaceae*	N+P

September/October

Ivy	*Hedera helix*	*Araliaceae*	N+P
Michaelmas daisy	*Aster novi-belgii*	*Asteraceae*	P

* - *Taraxacum officinale* is the common dandelion; *spp.* denotes many species, in this case many species of dandelion.

** - Top fruit include apple, pear, cherry, plum, etc.; all are *Rosaceae*.

For this Husbandry Examination the Examination Board have not made it clear whether it is necessary to know the botanical names and family of the plants concerned. We believe that some students and candidates will wish to have a deeper understanding, therefore we have included the full scientific nomenclature (genus, species and family) in the table above. It should be noted that the classifications of some plants change from time to time (eg. *Ragwort, Senecio jacobaea,* family *Compositae*. The family is now re-named *Asteraceae*. In Scotland it has a common name 'Stinkywilly').

It is extremely difficult to define what are the main plants throughout the country because there are many local variations and the student or candidate should be familiar with his local flora; eg. winter aconite *(Eranthis hyemalis* of the buttercup family *Ranunculaceae*) is a prime source of pollen in the spring in some parts of Devon and for the last few years our own bees have worked Lesser Celandine (*Ranunculus ficaria)* at this time of the year. When the authors lived in Sussex, neither of these two plants were worked by our bees in the spring because they were not available in that area.

It is to be noted that the labiates (eg. mint, thyme, rosemary, lavender, etc.) are all very attractive to bees but in UK they do not rate as major sources of forage (eg. thyme produces a crop of honey in Greece and Malta).

4.2 Any measures taken by the Candidate to enable the bees to forage on a particular crop and any special action needed as a result of foraging on local crops or a crop to which bees have been taken. Rape, Heather and Borage are three possible examples.

4.2.1 General.

Migratory beekeeping is implied in this part of the syllabus and the candidate should be familiar with the requirements of moving bees; reference should be made to Section 2.22 for moving colonies. Special preparation is required for working early spring crops such as rape and fruit for pollination contracts. Likewise, special preparation is required for stocks going to the heather later in the season; however, crops that bloom in the summer months such as borage will be managed normally for maximum population. Honey that granulates rapidly (within about two weeks of sealing) will require to be removed immediately the crop has bloomed and petal fall is in evidence. Each of these conditions is examined in the following paragraphs.

It is not clear to us why borage is given particular mention in the syllabus as this crop blooms in the summer and no special colony preparation is required in order to work it. It should be noted, however, that borage honey has a high sucrose content and has been specially listed in the Codex Alimentarius (2000) as a honey with a maximum permitted sucrose content of 15%. It is not clear at the time of writing whether this has been embodied in the Honey Regulations (which stipulate a maximum of 5%) and has become law in the UK.

4.2.2 Special preparation of colonies for:

   a) Working spring rape,
   b) Working heather,
   c) Pollination,
   d) Removing honey that granulates rapidly.

4.2.2.1 The preparation of colonies for a particular honey flow.

Special colony management is required for very early honey flows and those at the tail end of the season, for example, rape blooming in April and heather blooming in the last few weeks of August. All honey flows require the maximum bee population in each colony in order to take advantage of the available nectar; maximising the colony population for an April flow requires very early seasonal management not normally associated with the normal colony spring build up.

A typical season for both honey and pollination would start with plums followed by pears, cherries, apples, rape, sycamore, horse chestnut, strawberries, raspberries, beans, lupins, mustard, white clover, bramble and finally heather to name some of the major sources. The flowering period of some of the above will overlap and not all hives will be working the same crop at a particular time.

4.2.2.1.1 The basic requirements.

To manage colonies for maximum honey production there are some basic requirements which need to be observed irrespective of district, weather conditions or time of flowering of the forage plants. These requirements are discussed in the following paragraphs.

The starting point must be the seasonal colony population cycle (see Appendix 7). It is applicable in principle to our temperate zone in the United Kingdom and is valid for latitudes from about 35°N to about 65°N. It is based on a colony build up during the spring flow peaking in the summer (summer solstice) to take advantage of the main flow. It assumes that the colony does not swarm. The length of time that flows occur are quite long compared with higher latitudes. The graph of the colony population cycle is applicable to our native honeybee, another basic requirement where early flows are concerned. The yellow Italian bees work best in warm temperatures.

Listing the basic requirements we have:

- The type of bee.
- Good manipulative skills to prevent any swarming.
- Maintenance of disease free colonies.
- Queen selection for breeding purposes.
- Maintain all colonies virtually identical to enable them to be managed as identical units.
- All units must be as large as possible in bee population.

Very few beekeepers can meet the above requirements. Those who do may be compared to those gardeners with 'green fingers' whereby their plants always seem to succeed; in the beekeeping world the likes of Bro.Adam immediately come to mind. Such ability results in the colonies not swarming, colonies with maximum populations throughout the year which are then prepared to take advantage of any flow when it starts. Ideally the colony should be used in the most economic and cost effective way after the flow.

4.2.2.1.2 The bees.

Early flows means that the bees being used for these flows must be available early in the year in colonies as large as possible and they should be able to work at low temperatures. Bees from the tropics and sub-tropics, ie. the Mediterranean area, do not work in low temperatures and tend to have a slow colony build up in the chilly spring weather, waiting until it gets warmer before brood rearing starts with a vengeance.

It will be clear that a bee more suited to our climate is a prerequisite for this work and, of course, this means the black bee of European origin. Thus it is essential to be able to breed a bee suitable for the purpose and this has to be done from the stocks locally available to the beekeeper selecting for best honey yields and temper. The only way this can be done is by keeping good records and analysing them at the end of each year. Then it will be clear immediately which stocks are giving the hundredweights of honey and which are producing swarm cells in the spring.

Importing queens from other sources in the UK is no guarantee that the queens will be satisfactory and time will preclude experimentation in this direction as well as the expense involved. Unfortunately it has been our experience that the majority of beekeepers do not find the time for good queen rearing practice and while their bees produce average crops of honey, the bees are generally very bad tempered.

On the other hand, if early crops are not a pre-requisite then other races of honeybees can be entertained again with the objective of good temper and low swarming propensity.

4.2.2.1.3 The prevention of swarming.

If advantage is to be taken of a particular crop(s) during the active season then it is imperative that the colony does not swarm. If it does then about 50% of its population will be lost with a similar reduction in the foraging force. Reference to the graph in Appendix 6 makes this abundantly clear when compared with the colony that doesn't swarm (ie. the annual population cycle for a non-swarming colony). With the best will in the world a colony will make preparations for swarming, no matter how well it is managed and it is then incumbent on the beekeeper to prevent the loss of bees by adequate methods of swarm control.

4.2.2.1.4 Maintenance of disease free colonies.

Reference to Appendix 6, the graph shows the build up of a diseased colony which may be compared with a normal healthy colony (annual population cycle). It will be obvious that a diseased colony in the spring is likely to be a useless honey producer unless action is taken early in the year. This demonstrates the wisdom of monitoring all colonies for disease regularly; autumn and spring sampling for the adult bee diseases and at appropriate times for the brood diseases.

Whenever colonies of bees are moved they are put under stress, increasing the risk of weak colonies developing Nosema, EFB, etc., the well-known stress diseases associated with many commercial beekeeping enterprises. We do not think that there is a solution to this problem if the bees are continually on the move. Continual colony assessment can be the only answer coupled with hive hygiene but here it must be appreciated that this will increase the labour costs of the operation which are already likely to be running on a tight budget.

4.2.2.1.5 Selective queen rearing.

The provision of young queens in all stocks is a necessary preparation for a productive colony to make best use of the flow. Selective queen rearing is outside the scope of this syllabus but nevertheless candidates should take notice of this requirement; it is addressed in the modular examinations which follow this level of study.

4.2.2.1.6 Maintaining large stocks.

Maintaining large stocks clearly maximises the work force available to take advantage of any honey flow when it occurs during the season. In essence this means having a young prolific queen in a disease free colony which doesn't swarm, all of which have been discussed above.

4.2.2.1.7 Preparing the stock for an early flow in April, eg. rape.

The objective in preparing the stock for an early flow is to endeavour to have the largest colony possible by the time the rape comes into bloom and starts yielding nectar. Brood rearing has to be stimulated and to do this pollen is required. Brood rearing only starts with a vengeance when the foragers can collect fresh pollen. As brood rearing requires to be stimulated from mid January, when virtually no pollen is available, it is necessary to feed what are called, ' pollen patties'. One word of warning before preparation starts; do check with the grower which variety of rape is being grown as there are now types which produce virtually no nectar. We have had apiaries in recent years with rape the other side of the hedge and the bees have not collected a pound of honey.

4.2.2.1.7.1 Preparation of patties. See Section 2.19.2.

In order to build up the colony for the early spring rape flow the patties can be put on the colonies as early as the middle of January; the earlier the better to get the colony as strong as possible before the rape comes into bloom. The major drawback is that swarming problems are likely to occur earlier with such stimulation.

We have no experience of using pollen patties at other times of the year, eg. a flow associated with honeydew in July. There would be difficulties feeding the patties with supers on and we would not recommend feeding them at the entrance at that time of the year because of robbing problems.

4.2.2.1.8 Preparing the stock for a late flow in August, eg. heather.

Stocks destined for the heather in August have been used for the normal season up to and including the main flow. At this time of the year the strength of the colony is naturally declining quite rapidly, see Graph 1 in Appendix 6. It is normal practise to requeen the stock with a current year queen to encourage egg laying to continue for as long as possible to ensure maximum foraging force before moving the stock to the heather.

The brood chamber should be as full of brood as possible with food on the two end frames of the brood chamber to ensure that the colony does not starve if it is prevented from foraging during its first 10 days on the moor. Ideally the flow should be just starting when the stock arrives on the heather and, because the brood chamber is chock-a-block with stores and brood, honey is stored immediately in the super provided.

If the flow is good then the stock is likely to provide a surplus after providing enough winter stores for itself in the brood chamber

4.2.2.2 Working fruit blossom for pollination purposes.

Preparing stocks for pollinating fruit blossom entails an early build up very similar to that required for working autumn sown rape. The basic requirement in this case is that each stock shall have a maximum amount of open brood when actually taken to the fruit blossom requiring pollination. The reason for this is because the pheromone associated with open brood stimulates foragers to forage mainly for pollen rather than nectar. Generally, the pollination contract between the beekeeper and the fruit grower stipulates the minimum amount of brood (number of frames containing brood) required in each stock and any stocks can be inspected to ensure that the contract conditions prevail.

4.2.2.3 Removing honey that granulates rapidly.

It will be obvious that honey that granulates quickly should be removed from stocks after it is ripe and before granulation sets in. Usually with rape this is at petal fall as the fields are turning green. The danger to the stock is that it can be left with no food reserves of liquid stores particularly in May as the June gap starts to manifest itself. Feeding will be necessary to prevent starvation unless the colony has at least 10 pounds of stores at each weekly inspection, that is the amount a strong colony will use in 7 days foraging with little income as a result.

4.3 Honeydew, being able to name sources and describe the impact of honeydew in the area of the Candidate.

4.3.1 General.

Honeydew is the name given to the exudate of insects which feed on the sap (contents of the phloem tubes) of plants. These insects, are true bugs and belong to the order *Hemiptera* of which there are 50,000 species. 1,650 occur in the British Isles, all these bugs possess piercing mouthparts adapted for sucking juices of plants or other animals. There are three sub-orders *Coleorrhyncha, Heteroptera* and *Homoptera*. All the homopterans are plant feeders and the sub-orders contain serious pests. In the sub-order *Homoptera* there are two distinct divisions *Auchenorrhyncha* and *Sternorrhyncha*. It is the *Sternorrhyncha* that are the major producers of honeydew.

4.3.2 *Sternorrhyncha.*

These insects, mostly less than 5mm long, cause a vast amount of direct and indirect damage to plants by taking large amounts of sap, injecting toxins which may destroy chlorophyll, blocking the phloem tubes and transmitting viruses. Most of the bugs multiply rapidly in favourable conditions. Eggs overwinter and hatch the following spring, producing apterous (wingless), viviparous (producing living young), parthenogenic females. Several generations will be produced during warm weather, at the end of the summer male and female individuals will be produced, pair and produce eggs for the winter stage. The honeydew is exuded in small droplets from the anus during all the feeding stages of development but more especially during the larval stages. The exuded liquid is colourless, sweet and wholesome, it quickly solidifies in air and dries to a sticky, light brown, gummy like substance which will collect and retain pollen grains of wind pollinated plants, fungal spores, algae, soot and dust particles. The plant may develop a shiny black appearance due to the growth of the fungal spores on the honeydew. Honeydew may be produced so copiously that it can be seen falling from aphid infested trees. This may happen in urban areas where cars parked under trees will become covered in a sticky exudate.

The sub-order Sternorrhyncha is divided into five superfamilies, *Coccoidea, Aleyrodoidea, Psylloidea, Aphidoidea and Cicadoidea.* The manna of the bible story (Exodus Chapter 16 verse 3) is thought to have been the syrupy honeydew exuded by certain species of the family *Coccidae* (mealy bugs) belonging to the superfamily *Coccoidea.* In warm dry climates honey dew quickly solidifies into sugary lumps as the water evaporates and falls from the trees as *'manna'.* The bugs such as the aphids, greenfly and plant lice are especially characteristic of the northern temperate region belong to the superfamily *Aphidoidea.* There are over 3,600 species. They have a complicated life cycle and a highly developed polymorphism. The champion plant suckers are the aphids of which there are hundreds of species, 500 species occurring in U.K. The other three super families are of little importance to honeydew honey in temperate climates.

4.3.3 Honeydew and its sources.

Honeydew attracts honeybees when there is no other forage available usually between the months May to August. Many floral honeys contain a small percentage of honeydew. The honeydew also attracts ants, for whom the honeydew may be the main item of diet. Some species of ant have a permanent relationship with a number of the aphid species. The best producers of honeydew are always attended by ants. Honeybees collect the honeydew in the morning whilst it is still moist from dew. Later in the day it becomes sticky and gummy and more difficult to collect. This may be the cause of 'bad temper' reported whilst honeybees are collecting honeydew. In the northern temperate zone bees may collect honeydew from lime, oak, sycamore, beech, elm, ash, chestnut, hawthorn, fruit trees and various conifer trees. Modern farming techniques and the use of pesticides has greatly reduced the production of honeydew collected from aphids feeding on cereals and beans. Sections of honey, or cut comb can be spoilt by bees collecting honeydew.

In the U.K. the collection of honeydew by honeybees varies from year to year. Rex Sawyer reports in his book 'Honey Identification' that in the years 1970, 1976, 1983 and 1984 crops of red to brown honey containing honeydew were harvested in the UK. Pollen analysis of honeydew honeys may show grass pollen and other airborne pollen of non-nectiferous plants, characteristic square crystals, fungal spores, green algae, soot particles and mineral dust. Fifty or more years ago the lime trees in the cities were blamed for the sticky deposits which would fall on to the pavements or parked cars. In recent years either these trees have been removed, died from old age or been so severely pruned the honeydew is no longer a problem.

4.3.4 Definition of honeydew honey.

Honeydew may contain both the exudate of insects and the secretions of the extra-floral nectaries of a plant. It is usually collected when there is no floral nectar available. Honeybees collect both these sugary liquids and transform it into honey within the hive. 'Honeydew honey' means *'honey, the colour of which is light brown, greenish brown, black or any intermediate colour, produced wholly or mainly from the secretions of or found on living parts of plants other than blossoms.'* Definition taken from 'The Honey Regulations 1976'.

175

4.3.5 The constituents of honeydew honey.

As the basic raw material of the phloem tubes vary in content between species and at different times of the year so will the honeydew honeys vary in colour and taste eg. nitrogen content is high in the spring and low in autumn when the leaves change colour. The colour of the honey may vary from light brown, reddish or greenish brown to darker shades of brown. The taste may be anything from bitter, treacly, malty, like figs or toffee. The honeydew flow is determined by many complex factors including the population dynamics of the insects producing the honeydew.

Comparison of the constituents of average floral honey with honeydew honey

	Honeydew honey	Floral honey
Water	16-17%	18%
Fructose	1.8%	40%
Glucose	26.08%	35%
Other sugars	14%	4%
Undetermined	10.1%	
Ash	0.736%	(0.26%- 0.17%)
Nitrogen	0.10%	
Other substances		3%

4.3.6 Impact of honeydew in the foraging area of the Candidate's apiary.

Any candidate entering this examination is required to know whether there are any sources of honeydew in his area because if honeydew honey is sold then it must be in compliance with the Honey Regulations and labelled accordingly. Taking cognizance of the constituents of honeydew, in order to determine this beyond doubt would mean undertaking a pollen and chemical analysis of the honey concerned which is beyond the scope of this examination. Additionally, for the hobbyist beekeeper the cost of analysis is likely to be prohibitive.

Therefore, when the presence of honeydew in a particular apiary is suspected, great care should be taken when selling honey from that apiary in order to keep within the law.

4.4 Any sources of undesirable nectar found in the locality of the Candidate.

4.4.1 General.

Toxic substances and organic compounds are synthesised by various plants some with marked pharmacological activity . For example:

• Deadly Nightshade, *Atropa belladonna.* The juice of the berries causes dilation of the pupil of the eye.
• Foxglove, *Digitalis purpurea.* Digitalis can be extracted from its leaves; in its fresh state this can cause drowsiness, convulsions and even death. Where foxglove is grown extensively, honeybees

foraging for pollen have been poisoned.

• Common Ragwort, *Senecio jacobaea,* designated an injurious weed in the Weeds Act 1959, is poisonous to cattle and horses causing damage to the liver with *pyrrolizidine alkaloids.* Bees work ragwort blossom and obtain both pollen and nectar with no toxic effects. The honey is bright yellow with a strong unpleasant odour. After granulation much of the strong aroma disappears.

4.4.2 Some nectars, honeydews or pollens collected by honeybees are poisonous to the bees while others appear to be acceptable to bees but toxic to man or livestock. Those toxic to bees can contain toxins in the nectar or pollen or both. With toxic nectar bees die in front of the hive. Toxic pollen will affect brood and bees. Brood can die in the cell. Some toxins can affect the queen. 'Toxic' means poisonous and practically any substance can be toxic if taken in large enough quantities. There is a toxic level for almost any substance. Below this level, a material is safe, above it becomes harmful. Incidences of poisoning have been poorly documented. Sometimes abnormal environmental conditions produce effects that look like poisoning. As in cases of spray damage, the colonies affected should be fed sugar syrup in order to dilute the toxins and compensate for loss of foraging bees. The Authors understand that some French beekeepers feed a sample of honey to a dog as a test for toxicity before human consumption is contemplated.

4.4.3 The syllabus calls for undesirable nectars in the UK only. However, there are so few that we considered it desirable to provide a wider view of this restricted subject.

4.4.4 List of known plants which produce undesirable nectar and pollen or act as hosts to insects which produce toxic honeydew.

• In 1955 an incident of poisoning was reported in colonies of bees on the island of Colonsay off the west coast of Scotland. The bees had died out completely in 2-3 days after starting to collect nectar from Rhododendrum blossoms, *Rhododendrum thomsonii.* The death of the bees had been caused by the poison *andromedotoxin* or *acetylandromedol.*
• Other hearsay evidence includes reports of bees found dead or dying under lime trees, *Tilia petiolaris a*nd *T. tomentosa*, these incidents may have been due to the honeydew present on the trees or to the presence of mannose in the nectar. It has been reported that the casualties vary from season to season and affect the bumble bees more than the honeybee.
• Spring dwindling or '*May disease*' caused by the pollen from the *Ranunculaceae* family (buttercup) was reported in 1944 from Switzerland, probably due to the toxin *protoanemonin.* Nurse bees appeared at the hive entrance trembling and unable to fly, excitedly moving on the landing board, losing control of their legs, rotating violently on their backs, becoming paralysed and dying. The leaves of most species of buttercup are poisonous and avoided by livestock. The presence of *protoanemonin* in the sap gives the leaves a bitter taste. Each spring the stocks in the Authors' garden forage on the buttercup, *Ranunculus ficaria,* no ill effects have been observed over the last 10 years.
• The silver fir, *Abies alba,* is a source of honeydew which is toxic to bees. In Switzerland in 1951 thousands of returning foragers, with a waxy black appearance, were reported dying outside hives. Insects feeding on the plant sap of the silver fir convert the sap into simple sugars and synthesise new sugars by means of group transference eg. raffinose, rhamnose, stachynose, melibiose and

mannose. Some of these sugars may be the cause of toxicity in bees foraging for honeydew which varies in composition depending on the species of sap-sucking insect present on the tree.

• In the United States of America the 'California buckeye' chestnut tree, *Aesculus californica*, produces poisonous nectar and pollen. The bees become black and shiny, trembling and paralysed. Non-laying queens, dying brood and infertile eggs have also been reported. As this species covers 14 million acres in North America its effects on honeybees are well known to local beekeepers. The mountain laurel, *Kalmia latifolia* found in the Appalachian Mountains contains the same toxins as Rhododendron. *Andromedotoxin* accumulates in honey and can be poisonous to man, bees and cattle.

• In Asia Minor, the earliest account of poisonous honey was recorded by Xenophen in the year 40 BC. It was thought that his troops were poisoned by eating the honey from hives of bees that had been foraging on *Rhododendron ponticum*. Around Trebizond today bees are kept not for honey but for wax production.

• New Zealand, where bees were introduced to pollinate the clover crops, has two trees which are dangerous to bees or mankind. The Karaka tree, *Corynocarpus laevigatus* produces flowers which are very attractive to bees. The nectar probably contains *karakin* which poisons the foraging bees, the colonies die out due to loss of foragers. The Tutu tree, *Coriaria arborea i*n Northern New Zealand does not yield nectar but bees collect pollen and honeydew exuded by the leaf hopper, *Scolypopa australis*. The honey produced contains *picrotoxin tutin* which was thought to be responsible for an outbreak of poisoning amongst local inhabitants, the honeydew was not toxic to bees.

4.4.5 Conclusion.
We believe that the chances of bees collecting undesirable nectar, or pollen for that matter, are extremely remote and the majority of beekeepers will never encounter problems in this direction.

Genetically modified crops are being grown in the UK. There has been much written in the national and beekeeping press about these experimental plantings, mainly maize and rape crops. The questions which remain unanswered are:

• Will these genetically modified crops be detrimental to animal or human health?
• Will pollen and nectar collected by the bees be detrimental to the health of the honeybee?
• Will honey be contaminated by this genetically altered pollen?
• Will wild flowers be cross-fertilised by this genetically altered pollen?
• Will the wild life, insects, birds, butterflies, etc. which feed on the genetically altered plants be affected?
• Will the produce of organic farms be contaminated by areas of genetically modified crops?

** ** ** ** **

DISEASE, PESTS AND POISONING

The Candidate will be able to:

5.1 Describe how disease can be spread between colonies and how good management practice can reduce disease occurring.

5.1.1 How disease is spread between colonies.

It is spread as follows:

a) By robbing; likely when the infected colony becomes weak and is then robbed by strong healthy colonies either within an apiary or between apiaries.
b) By drifting; adjacent hives in an apiary which are not orientated in different directions.
c) By feeding infected honey; bees should always be fed sugar syrup.
d) By bees gaining access to infected honey, combs, wax and propolis left around the apiary or within foraging distance.
e) From appliances (eg. hive tools, extractors, etc.).
f) From second hand infected equipment (combs, hive parts, etc.).
g) From swarms of unknown origin.
h) By the exchange of combs between hives.
i) By the purchase of bees from a doubtful source.

5.1.2 How good management practice can reduce disease occurring.

Most of us have been brought up and educated, from our Basic Examination, to observe good apiary hygiene which is summed up in the following, often referred to as the 10 commandments of good beekeeping:

1. Always keep the apiary clean and tidy.
2. Never throw propolis or brace comb on the ground; be sure always to place it in a suitable container and remove it from the apiary.
3. Never buy old combs.
4. Never buy colonies of bees unless it is known that they come from disease free apiaries; never accept stray swarms from unknown origins.
5. Always disinfect second hand hives and other equipment before use.
6. Never feed honey or allow bees to gain access to it; refined sugar is the only acceptable feed for honeybees.
7. If a colony dies out during the winter (or at any other time) and the trouble is not due to starvation, seal the hive, pending the examination of a sample comb and bees, to prevent the remaining stores being robbed out.
8. Never exchange brood or super frames/combs between one colony and another unless it is known that all colonies are free from disease. Where possible, supers should be marked and always used on the same colonies.
9. Take care to prevent robbing at all times by observing item 2 and not spilling syrup or having

leaky feeders.

10. Arrange all hives in such a way that drifting is reduced to a minimum.

The code is essentially for minimising disease and its spread in an apiary and appears in a modified form in ADAS leaflet P306 - 'Foul brood of bees: recognition and control'.

5.1.3 The spread of disease between colonies in different apiaries.

This is a very real danger which penalises the good beekeeper who maintains very strong disease free colonies. Bees from these colonies rob weak colonies whether they be in the same apiary or in one some distance away. Unfortunately there is nothing that the beekeeper can do about it except to be vigilant at every colony inspection to look for signs of all the known diseases.

5.2 Describe the signs of American Foulbrood.

5.2.1 Field diagnosis of AFB and EFB.

ADAS leaflet P306 Revised 1982 'Foul brood of bees: recognition and control' should be obtained which contains some excellent coloured photographs of both brood diseases.

• Both diseases are diseases of the brood and there are no signs associated with the adult bees in an infected colony. In order to diagnose them in the field, it is necessary to open up the colony and examine the combs containing brood. To do this properly it is necessary to shake the bees off the comb before examining it, leaving no more than a few bees on the comb. The reason for this is that in the early stages only an odd cell or two will be exhibiting the tell-tale signs. This important aspect of searching for the diseases is frequently overlooked and inadequately expressed in much of the literature. There is a right and a wrong way of shaking bees off combs, the objective is to rid the comb of bees and keep them in the hive (not flying around the apiary); therefore raise the comb slightly by holding the frame by the lugs and shake it sharply within the brood chamber without jarring the rest of the colony.

• In order to diagnose the diseases in the field it is easier to remember the signs if one has an understanding of the progress of the diseases.

5.2.2 Progress of AFB (American Foul Brood).

The larva is fed the AFB spores (*Paenibacillus larvae larvae*) with the larval food. The spores germinate in the ventriculus and the larva dies after the cell is sealed. The germinated spores break through the wall of the ventriculus into the haemolymph and the larva dies of septicaemia * ; then the whole larval form disintegrates, melts down, becomes thick and sticky and finally dries to a hard scale on the lower angle of the cell. During this deathly saga the colour changes from white to black. It is most important to note that prior to the sealing of the cell, the larvae appear to be perfectly healthy.

   * Septicaemia - is the circulation and multiplication of micro-organisms in the blood.

5.2.3 Signs of AFB.

• open brood - no signs,
• sealed brood, many signs as follows:

1. After the larva dies, the domed cell cappings become moist and darken in colour.
2. Cappings then sink and become concave (still moist and discoloured).
3. Holes appear in the cappings (ie. perforated).
4. Matchstick pushed through sunken capping to test for roping of the contents. Length of 'rope' between 1 and 2 cm. This roping is considered to be a positive identification of the disease. Colour of cell deposit is from light brown to nearly black. The roping test can only be done between the time the larva has 'melted' and the remains thickened slightly and before it has dried too much to be sticky.
5. The remains dry out on the lower angle of the cell and form a hard black scale. By the time the scale is formed, the bees have uncapped the cell completely and tried to remove the scale. In order to see the scale the comb must be held at an angle with the top bar closest to you and the bottom of the frame away from the body (angle about 45° to the vertical). Good light is essential, some books say from the back while others say from the front; we think either is acceptable depending on whether you are in or outdoors. In the early stages of the infection, possibly only one or two cells may have scale and this is why it is so important to clear the frames of all the bees when doing an inspection for foul brood.
6. Brood combs which have a 'pepperpot' appearance (ie. empty cells among sealed brood) should be treated with suspicion and examined closely for any sign of scale.

• AFB infections have no smell; many books indicate a foul odour. *Paenibacillus larvae larvae* when sporulating releases an antibiotic preventing any secondary infections. If an offensive smell is present it will be due to secondary infections or some other cause or the confusion may arise because when the bacteria are in the rod form all the cells are sealed and no odour can be released. Rely on visual signs, not odour, for AFB.
• AFB is easily identified visually in the field, however, it can be confirmed if necessary by laboratory tests usually on a piece of scale from the comb.

5.2.4 Other points of interest.

• The signs are well documented but there are one or two areas where rational explanations are missing. It is well known that the cappings of the cells sink after the larvae die. Why? As the cappings are porous to allow the larvae to breathe, the sinking cannot be due to a change in pressure. We have not been able to find a reasonable explanation for this phenomenon.
• There is a dilemma about smell. Some authorities say AFB has a characteristic fish-glue like smell and others say that there is no smell. Smell occurs during the putrefaction of most animal tissue and as the larvae putrefy it would seem reasonable that a smell is produced. The ingested spores (1.3 x 0.6μm) germinate and start to multiply as rod shaped bacteria (2.5 to 5μm x 0.5 to 0.8μm) and being motile penetrate the gut walls and continue multiplying in the haemolymph until the larvae die of septicaemia. The bacteria stop multiplying and start to sporulate about the same time that putrefaction starts. Secondary organisms are unable to grow due to the antibiotic released and it is the secondary organisms which cause the smell. In the absence of further information on this point we consider that smell is not a good sign for the diagnosis of AFB.

• The most characteristic sign of the disease is that the pupal proboscis protrudes from the scale to the centre of the cell, something which never occurs in other brood diseases.

• It is possible to have both AFB and EFB on the same comb and in the same hive. As some of the secondary infections associated with EFB smell, then confusion can occur.

• Dry scales fluoresce strongly in UV light. As small battery operated UV lights are becoming quite common for security marking, we believe they could be a useful diagnostic tool but we have not tried them out. We have the lamp but not the AFB!

• There are no signs of AFB in open brood; larvae die after the cell is sealed and all signs of the disease are of the sealed brood or much later scale which the bees cannot remove. However, in the very early stages of the infection a suspicious open cell or two may be found where the larvae appears to be 'not right'. Such larvae should be tested for roping. When only a few cells are infected initially the bees open them up and reveal the dead larvae which is usually honey coloured. We saw a demonstration by the SW RBI, Len Davie, showing how to examine an apiary for foul brood. Out of six colonies in the apiary, he found one cell in one colony which was open and a matchstick demonstrated roping of the cell contents. This was a most excellent demonstration of the expertise we should all be striving to attain.

5.3 Describe the signs of European Foulbrood.

5.3.1 Field diagnosis of EFB.

See sub-section 5.2.1 above. As all foul brood diagnosis commences in the field it is very important to be able to shake the bees off the frames adroitly with a minimum of fuss.

5.3.2 Progress of EFB (European Foul Brood).

The larva is fed the pathogen (*Melissococcus plutonius*), this time a bacteria which is not spore forming as was AFB, which multiplies in the ventriculus by using the larval food and the larva dies before the cell is sealed due to starvation. It dies at about day 3 or 4 before the cell is sealed, so it is quite large when it dies. A dead larva is not sealed by the bees and is removed. During the starvation period the larva contorts into unnatural shapes in its cell and changes colour from a pearly white to cream to yellow to light browny green (colours are difficult to describe in words; any deviation from the pearly shiny white must be regarded with suspicion). When the bees remove the dead larvae, they are removed in one piece. The infected larvae are either there to see or else the signs have been removed.

5.3.3 Signs of EFB.

• Sealed brood - no signs (the larva dies before sealing).
• Open brood - the signs are as follows:
    1. Larvae are usually in unnatural contorted positions in the cells; twisted spirally or flattened out lengthwise (nb. stomach ache is a good analogy).
    2. The colour changes from a pearly white of a healthy larva to dull cream, to light brown and eventually a greeny hue. The colour change should be associated with the unnatural positions.

3. The dead larvae have a melted down appearance but still have a larvae-like shape.

4. Again EFB does not itself smell. However, very often an offensive smell is present on combs with EFB infected larvae; these are secondary infections often associated with EFB and are another indication that the disease may be present (3 secondary infections are due to *Bacillus alvei* which is very common and has a foul odour, *Bacterium eurydice* which is most common, *Streptococcus faecalis* which is fairly common causing a sour smell and finally *Bacillus laterosporus* which occurs occasionally).

• EFB is very difficult to diagnose positively in the field for the following reasons:

1. The larvae are removed quickly from the hive once they are dead so the evidence is often removed and not there for the beekeeper to see.

2. Any diseased larvae can be confused with other brood diseases, such as Sacbrood or Neglected drone brood, unless the beekeeper is very experienced. Most Foul Brood Officers will remove a frame with dead and dying larvae for laboratory analysis.

3. The best time to look for EFB is when the brood outnumbers the adult bees (see colony population cycle - Appendix 8) in the spring about mid April to early May. At this time the chances of spotting the diseased larvae are greater because the house bees are 'fully stretched' under these conditions.

5.3.4 Other points of interest.

• The original name *Bacillus pluton* was changed to *Streptococcus pluton*, later changed again to *Melissococcus pluton* and recently corrected to *Melissococcus plutonius*.

• The confusion in the names arose due to an original error in classification; the genus Bacillus is spore forming and EFB is a non-spore forming bacterium. Then the guanine and cytosine content of its nucleic acid were determined, which excluded it from the genus Streptococcus and it was re-named and became the type species of a new genus Melissococcus.

• This disease of the brood suddenly appears and just as suddenly disappears during the active season in early summer 'now you see it now you don't'. There is mounting evidence that EFB could well be endemic in UK colonies and research work is being undertaken at the time of writing.

• Again there is confusion about the smell; for example, Bailey states when discussing dead larvae "..turn brown and decompose, often giving off a foul odour or a sour smell, but sometimes having little or no smell". We believe that smell is an unreliable sign of the disease and visual signs are more reliable.

• To the untrained eye EFB can be confused with Sacbrood and it is always essential to undertake microscopical analysis to determine whether the disease is present. The most reliable way of undertaking a bacteriological examination is to select a larva with a white mid-gut and pull it apart on a slide revealing the bacteria in white clumps. A healthy or lightly infected larva will have its mid-gut a golden brown colour.

• Larvae that die decompose rapidly and if left in the cells quickly dry to a brown scale which is easily removed by the bees.

• If larvae are infected with *M. plutonius* and are capped and manage to pupate, they form either normal adults or undersized adults and leave infective *M. plutonius* in their faecal remains in the cell when it is vacated. This is how the disease is spread within the colony and undersized adult bees can be a sign of EFB.

• Because the bees remove the dead larvae then a pepper pot brood pattern must always be a possible sign of EFB as well as other bee diseases.

5.4 Describe what actions shall be taken to comply with statutory requirements if a brood disease is suspected.

It is important that candidates for the examinations obtain and study "The Bee Diseases Control Order 1982, S.I.107 (AFB, EFB, Varroasis)" obtainable from H.M. Stationery Office.

5.4.1 Statutory Instrument 107.

S.I.107 was well drafted and is summarised briefly below:

It deals with 3 diseases namely, AFB, EFB and Varroosis (which we will ignore here).
If disease is suspected the owner **must notify** it and **must not remove** the suspected colony until:
    a) an authorised person has examined the colony,
    b) a report is received confirming that disease is not present,
    c) notice has been served removing prohibition.

The authorised person may:

    a) take samples of bees, combs, etc.,
    b) mark any hive or appliance.

If the authorised person suspects disease he shall:
    a) serve a notice prohibiting removal,
    b) send samples for analysis with all speed.

A report to be sent to the owner withdrawing prohibition if disease free. If the report confirms disease or the owner agrees with the authorised person that disease is present and signs a copy of the notice, the Minister may:

For AFB:
a) give notice to destroy by fire or treat as specified under supervision.
b) notice (standstill order) to remain in force until cancelled.

For EFB:
a) give notice to destroy by fire or treat as specified under supervision.
b) period of 8 weeks to remain in custody after treatment except by special licence.
c) notice to remain in force until cancelled.

For Varroosis:
Most of the procedures have been changed and the agreed plan for the first discovery of Varroosis went by the board for reasons best known to the Ministry. Candidates should study and be aware of the latest procedures.

5.4.2 The implementation of the regulations.

All regulations in respect of bees, both for importation and the control of notifiable diseases are vested in the Central Science Laboratory (CSL) an agency of DEFRA (Department of Environment, Food & Rural Affairs) formerly MAFF which initially came under the auspices of ADAS. The organisation for the control of notifiable diseases, set up during 1995, provides 10 Regional Bee Inspectors (RBI) to cover the whole of England and Wales. It is too soon to comment on how well this reorganisation will work and how effective it will be. The declared objectives are that the RBIs shall pass their knowledge on to the beekeeping organisations in their regions so that they become self-reliant and the need for RBIs will eventually become unnecessary. We believe this to be unrealistic and self-policing of statutory regulations can never work satisfactorily.

5.4.3 The action to be taken when AFB or EFB is found.

5.4.3.1 Two avenues exist:
   a) when found by the beekeeper - report findings to DEFRA immediately and follow their instructions,
   b) when suspected by the beekeeper and found by DEFRA (FBO) - follow their instructions implicitly.

If AFB is diagnosed then the treatment is always to destroy the colony; the beekeeper being served a notice which he must sign. If EFB is diagnosed then the condition may be treated with antibiotics but only at the discretion of the FBO. Individual beekeepers are not allowed to treat their own colonies and it is against the law to do so. The work of destroying a colony falls to the owner but must be supervised by the FBO. The FBO usually administers any antibiotic treatment for the owner which is followed up at a later date by further inspection of the colony and, more than likely, any other colonies in the apiary and or adjacent apiaries. In both diseases a standstill order is put into effect banning the movement of colonies and equipment into or out of the apiary concerned. The standstill order is operative until such time as it is cancelled by the FBO in writing.

5.4.3.2 Colony destruction treatment for AFB:

• Colony may only be destroyed after dark when all bees of that colony have returned and stopped flying.
• Before the evening seal all openings (except the entrance which is reduced to c. 2") and put zinc gauze over the feed hole.
• When the bees have stopped flying, block the entrance securely (eg. clod of earth) and then pour ½ pint petrol into the colony through the feed hole. The bees will be dead in a few minutes. Leave for 10 min.
• Dig hole 3 ft. square by 3 ft. deep. This can be done earlier and in a position not too far away to minimise carrying equipment to the burning site at the bottom of the hole.
• Prepare starter paper + 2/3 combs and frames pyramid fashion.
• Set alight and burn all hive contents including combs, frames and quilts if these are used.

• Scrape all boxes, floorboard and crown board free of wax and propolis, all scrapings put into the fire.
• When completely burnt out back fill immediately.
• Scorch all hive parts with a blow lamp to a coffee brown colour paying particular attention to corners and cracks in the woodwork and to the queen excluder.
• Disinfect any other appliances eg. smoker, feeder, hive tool, etc. in solution of:
  - 1lb washing soda,
  - ½lb bleaching powder,
  - 1gallon warm water.
Use while warm and rinse in clean water before drying (note that the solution is caustic - take care).
• Finally obtain a destruction certificate from the FBO to substantiate your insurance claim on BDI.

5.5 Describe how to distinguish between female Varroa and female Braula coeca.

5.5.1 General.

The syllabus requires the candidate to distinguish between the females of the species only because they are the ones that are likely to be seen within a colony. The ability to be able to spot the differences between the male and the females of the species concerned is beyond the scope of this examination. It should also be noted that there is no such animal as 'Varroa', it may be called a varroa mite, however, the proper name is *Varroa destructor* formerly *Varroa jacobsoni*.

5.5.2 *Braula coeca.*

All candidates must be aware that *Braula coeca* is neither a mite nor a parasite, but is an insect that steals food. *Braula coeca* (Nitzch) in the family *Braulidae* is called a bee louse but in effect it is a wingless fly and not a true louse. The family is contained in the order *Diptera* (true flies). The two true lice orders are *Anoplura* (sucking lice) and *Mallophaga* (biting lice). The relationship between *Braulidae* and other *Diptera* is still enigmatic and discussion continues about the placement of *Braulidae* in the animal kingdom. All adult *Braulidae* lack halteres and wings.

All *Braula coeca* are reported to be inquiline (an animal living in the home of another animal) and none are reported to be harmful to the colony. The life cycle is as follows:

• Eggs are laid on the inner side of honey cappings and sometimes on the wall of cells filled with honey.
• Eggs hatch to larvae which feed on wax and pollen (found in their intestines) forming ever-lengthening tunnels in the wax cappings which widen as the larvae grow.
• The larvae pupate at the wide end of the tunnels and finally emerge as adults.
• At emergence the adult *B.coeca* is white and changes to its characteristic reddish / brown colour in 12 hours as its exoskeleton hardens.
• Development is entirely under the cappings of honey cells and it is not associated with brood cells in any way.

- Adults mainly inhabit the petiole of worker bees, queens and occasionally drones. They move to the mouthparts of the bee when it starts to feed.
- When the bee is feeding, the *B.coeca* resides on the open mandibles and labium reaching into the cavity at the base of the extended glossa near the opening of the duct of the salivary glands. It is not clear whether it feeds on the salivary gland secretions but it is thought to be highly likely.
- The queen appears to be more attractive than the other castes and the maximum number on a queen has been reported as 30 at one time. The maximum daily collection from a queen is reported as 371! Some authorities consider that infestation at these levels would have some effect on reducing the laying ability of a queen.
- Breeding takes place between May and September and the louse has the ability to overwinter with the bee.

No ill effects on the colony have ever been reported but damage to cappings of honey sections and cut comb honey from the larval tunnels occurs. It is always advisable to put prepared comb for sale into the freezer (-15°C) for a couple of days to kill any eggs and larva. It is seldom necessary to specifically treat a colony for Braula infestation. It is now becoming evident that all chemical treatments for Varroosis are lethal to the *Braula coeca* and it could possibly become an 'endangered' species.

5.5.3 The differences between *Varroa destructor* and *Braula coeca*.

All detection for Varroosis is dependent on being able to recognise a *Varroa destructor* and not to confuse it with a *Braula coeca*. The physical differences are as follows:

Braula coeca: ellipse shaped c.1 - 2mm with 6 legs, coloured reddish brown. It is a wingless fly. Initially it is white and takes about 12 hours from hatching to develop its colour. The head and posterior end of its abdomen are on the ends of the major axis of the ellipse, the legs are on the sides associated with the ends of the minor axis of the ellipse looking down on the dorsal side. Easily seen by eye riding on worker bees and very often the queen is infested. Causes no harm to queen or bees but the larvae spoil honey comb with fine tunnels in the cappings.

Varroa destructor: also ellipse shaped c. 1.1 - 1.7mm. with 8 legs, coloured reddish brown the same as the Braula coeca. The legs are on the ventral side and cannot be seen when it is viewed looking down on the dorsal side. It travels 'blunt end first', its legs being on the sides associated with the ends of the major axis. It is an arachnida and is in the spider class in the animal kingdom not the insect class. These mites are difficult to detect as they feed on the haemolymph by piercing the membrane between the abdominal segments on the adult bee and breed in the capped brood cells.

5.6 Describe the method adopted in the Candidate's apiary to monitor and control Varroosis.

5.6.1 Methods for detecting the presence of *Varroa destructor*.

Varroa destructor is a parasitic mite which lives on the exoskeleton of the honeybee and breeds in the sealed cells of worker and drone brood. All methods of detecting the mite are visual. There are now a wide number of methods for detecting *Varroa destructor* infested colonies but for the purpose of this section only three simple methods will be described which are seasonal in use.

a) Examination of hive debris on the floorboard in early spring.
This method has been advocated by the NBU for some years and at the time of writing they will still undertake the examination free of charge. The method has its limitations and is not wholly reliable particularly when the infestation in the colony is light. The debris from all the floorboards in an apiary is collected and bundled together and sent to the NBU with the name and address of the beekeeper and his apiary. After examination, the NBU send back a report stating the results. If the sample comes back with a negative result and a light infestation has been missed it will be of little consequence on colony performance during the coming season. The method must be quite attractive to many beekeepers until such times as it is discontinued or until the apiary concerned has been diagnosed positive.

b) Uncapping brood.
This is used particularly on drone brood at the pink eye stage of development, using an uncapping fork to expose the drone pupae during regular colony inspections. Care must be taken when examining the larvae for reddish coloured mites; in the early stages of development the nymphs are virtually colourless and translucent. Special frames of drone foundation can be inserted in the brood chamber for this purpose. Note that this is also a manipulative method for the control of Varroosis without chemicals.

c) Bayvarol/Apistan test using a varroa screen.
The varroa screen must be used in conjunction with a paper insert below together with one strip of Bayvarol manufactured by Bayer or Apistan made by Sandoz. It is prudent to smear a thin layer of grease around the edges of the paper insert to prevent any mites that are knocked down and not killed from walking back into the hive again. Bayvarol and Apistan are the only medicaments approved for use in the UK at the time of writing. Bayvarol strips contains 3.5mg of flumethrin and Apistan strips contain 8g of fluvalinate, the test strip should be left in the colony for 24 to 48 hours. The strip and insert are then removed and the insert examined with a magnifying glass for dead mites. The strip should be inserted between the centre frames of the cluster and can be introduced without moving the frames and disturbing the colony. It is the most effective detection method available at the present time. The strip can be re-used to test other colonies in the same apiary, noting that other diseases can be transferred from one colony to another on the same strip. When Bayvarol is used for treatment the strips are left in the colony for 5 to 6 weeks, ie. the strips have a working life, according to Bayer, of 6 weeks using 4 strips in an average size colony. For testing purposes it would be prudent to use the single testing strip in the colonies for say no longer than 21 to 30 days of actual use before it is scrapped thus ensuring a reasonable surface density of flumethrin to knock down any mites that may be present. The test can be used at any time but is usually applied in the spring and autumn. Only 2 strips of Apistan are required to treat an average size colony.

5.6.2 Methods for monitoring the mite level in a colony throughout the year.

Once the mite has been detected in the colonies of an apiary, any monitoring process must involve counting dead mites that have died a natural death or counting the mites killed by knock down tests at regular intervals. If mites have been detected in one colony in an apiary, then it is safe to assume that all colonies in that apiary are infested or will be very shortly even if tests on these other colonies prove negative in the initial stages.

To monitor dead mites it is necessary to have a special floorboard with an integral Varroa screen (which are unpopular because of their high cost) or resort to a DIY job with the existing floorboard turned through 180° with a screen above.

The frequency of each monitoring session must be decided and the more often the better for the best results. Whether this is once a week or once a month must rest with the beekeeper and the amount of time that he can devote to the monitoring as part of his management system. Once the number of colonies rises say above five, the work of monitoring will quickly become very time consuming and tedious.

It is not much use counting mites as a continual monitoring process unless action is taken if the levels rise above a predetermined threshold. For management purposes it is not clear what this level should be. It should also be noted that if the colony has had supers added then it would be unwise to treat until they are removed which is likely to be at the end of the season. The question must therefore be asked whether there is any purpose in monitoring on a continuous basis throughout the season?

We are of the opinion that any monitoring should be regarded, at present, as experimental in order to determine possible management methods for the future. The equipment required to provide continuous monitoring demands screens that do not corrode and are made of a material that cannot be damaged by the bees themselves.

It is highly unlikely that bee farmers with commercial enterprises will have sufficient time or effort available to undertake monitoring programmes; the economics of running such businesses would preclude it. Most of the hobbyist beekeepers are unlikely to be interested also. Thus any monitoring is likely to be left to the informed hobbyist beekeeper who has an interest in this new disease. Thus the majority of beekeepers in the UK will require control methods to be applied annually or bi-annually that are simple to use and relatively cheap.

During 1998 the CSL produced a calculator (based on work undertaken by Dr.Martin) which takes into account the natural mite drop, the month of the year the mite drop is measured and then predicts whether to treat or how many days grace the beekeeper has before treatment is necessary. It has not found much favour among beekeepers and it is not clear how effective it is as a management tool. Our own experience with it is not at all favourable.

Until more recommended methods of monitoring and treatment are enunciated the beekeeper is very much on his own and we advise ringing the changes with respect to treatment from Apistan and Bayvarol, manipulative to essential oils and various acids (to be used with great care).

5.7 Discuss the impact of virus damage related to Varroosis.

189

5.7.1 The effects of the mite infestation on the colony.

It is instructive to examine the effects of the parasite on the individual worker bee in the colony:

The number of mites per cell (or per bee) has a marked effect on the haemolymph of the bee, the protein content being reduced by 15 to 50%. Brenda Ball quotes 1 to 3 mites cause a 27% reduction and 4 to 6 mites a 50% reduction.

The protein reduction in turn results in a marked reduction in the final weight of the bee (6 to 25 % weight loss) and a reduction in the longevity of about 34% to 68% of the adult life span.

If 5 or more mites are present then there is a high probability that the bee will be killed. If not there will be marked damage to the wings, legs and abdomen. Infestation must be at an advanced stage and the visible signs such as crawling and deformed bees at the entrance will only represent a very small fraction of the colony damage.

The morphological changes are small when bees are infested with 1 or 2 mites (eg. 1 to 3% reduction in wing length). This demonstrates the importance of early detection in the management of colonies because it is so difficult to detect any damage.

It takes approximately 3 to 5 years before the colony is weakened, that is, when the mite population = 30 to 40% of the adult bee population (eg. 40,000 bees and 12,000 mites). At this stage there will be a rapid decline in the adult bee population with severe brood damage, reminiscent of the signs of EFB. The death of the colony will quickly follow.

If the infestation reaches a level of 1 mite / cell and if treatment is not undertaken the colony will die out in 2 to 4 years. At the terminal stage the collapse is very rapid.

It will be clear that due to the general weakening of the bees and the colony coupled with the reduction in longevity, the normal house bee and nurse bee duties will be seriously disrupted leading to poor colony hygiene and a deterioration in the hive environment.

It is clear from the above that the signs of Varroosis are virtually impossible to detect visually until it is too late. Early diagnosis is essential for good bee husbandry.

5.7.2 Virus damage due to Varroosis.

It is unlikely that any sign of Varroosis will be apparent until the colony has been infected for about 3 years. The first indications are likely to be a general weakening of the colony.

As the *V. destructor* breeds in the sealed brood cell (drone preferred) these also cannot be seen except by opening the cells and conducting a systematic search. At an advanced stage of infestation

(say 2 years +), when the infestation has become heavy, underweight workers will be produced due to malnutrition and then deformed worker bees are likely to appear in the colony. Such bees may have stunted bodies and / or deformed bodies, wings and legs. There are some good colour photographs in the MAFF 1992 pamphlet (PB 0925 on the back page) of these deformities. There are likely to be 3 or more mites per cell at this stage and there will also be signs of neglected brood and the spread of secondary infections including viral infections.

When the colony is at an advanced stage of infestation there is every likelihood that it will collapse suddenly and die out most probably at the end of the summer. The suggested mechanism is that as the queen reduces her laying after the main flow there are fewer brood cells for the female mite to enter and more than one mite per cell results. With multiple females occupying one worker cell deformed bees are inevitable leading to rapid colony collapse. This emphasises the importance of detecting the mite at an early stage of infestation in order that the colony may be saved in time.

The viral diseases are responsible for the demise of most bee colonies and not the other diseases with which they are associated. The exceptions are the two foul broods. The other diseases and/or mite infestations debilitate the colony putting the bees under stress and allowing the viral infections to take over from, for example, Nosema, Acarine, etc.

Many of the viral infections of the honeybee are associated only with its adult form such as Chronic Bee Paralysis Virus (CBPV), Acute Bee Paralysis Virus (ABPV), Black Queen Cell Virus, Bee Virus X, Bee Virus Y, Slow Bee Paralysis Virus (SBPV), Filamentous Virus, etc. They could be classed as adult bee diseases. Conversely there are few viral infections associated with brood diseases such as Sacbrood Virus.

  • Viral diseases shown to be associated with Varroa infestation are BQCV, BVX, BVY, ABPV (Germany only), CBPV (Germany mainly, low in UK), SBPV (recently in UK).

5.7.3 Slow Paralysis Virus.

By injection it causes death after about 12 days. In recent investigations in the UK, SPV has unexpectedly increased in prevalence and has caused both adult bee and brood mortality in Varroa mite infested colonies.

5.8 Discuss the impact of re-infestation of Varroa on the management and timing of Varroa control.

5.8.1 General.

This is a new topic in the BBKA series of examinations and very little has been written about the subject. It is thought to have arisen as a result of the BBKA initiative of encouraging beekeepers to treat for Varroosis on a national basis at a set time each year in order to minimise the total number of mites at the end of the season prior to colonies settling down for winter. The main flow is complete by the end of July. Allowing 2 weeks to remove the crop, it has been recommended that the treatment for Varroosis starts in mid August and should be complete by about the end of September.

5.8.2 Re-infestation.

Varroa mites do not fly and the only way re-infestation can take place is by bees drifting from one colony to another within an apiary (note that drones are notorious for drifting from colony to colony in the same apiary but only during the summer) or by bees robbing either within the same apiary or between two apiaries. It has been suggested that after treatment, re-infestation can occur from bees owned by a beekeeper who has been antisocial enough not to treat his bees at the appointed time. This could only occur if robbing takes place, the robbers coming from a strong colony that has been treated ready for winter. For the sake of this discussion bees taken to the heather are excluded as they will be treated separately when they return.

5.8.3 Conclusion.

Generally, after the end of September into October bees will be foraging on the last few plants of the year such as ivy and it our experience that there is virtually no robbing between colonies at this time of the year with reduced entrance blocks in place and guards in evidence at the entrances to most colonies. Furthermore, Varroosis has now been established as an endemic disease since 1992 and those irresponsible beekeepers who did not treat their bees have lost them and many of these have ceased to keep bees. There are one or two isolated cases where they collect swarms, take whatever crop they can from them and lose them again as a result of failing to treat them; these are the jokers who cause the trouble because their stocks are likely to be weak and easily robbed. Thankfully there are few of them.

On this basis we consider re-infestation, if indeed it does occur, is very unlikely and is of little consequence to the average beekeeper.

5.9 Describe the impact of Nosema disease on a honeybee colony, and its diagnosis and treatment.

5.9.1 General.

Nosema - causative agent - *Nosema apis* (Zander) - a spore-forming protozoa.

• The Nosema spore (6 to 8μm) can only be observed using a compound microscope. It was discovered by Prof. Enoch Zander at Erlangen in the early part of the century. The protozoa multiply in the ventriculus (gut) of the honeybee (30 to 50 million spores when infection fully developed) and impair the digestion of pollen thereby shortening the life of the bee. It does not affect the honeybee larvae.
• Signs and effect:
　　1. Infected bees themselves show no outward signs of the disease.
　　2. Colonies fail to build up normally in the spring (see Appendix 6).
　　3. Badly infected colonies in the early part of the year:
　　　　- exhibit signs of dysentery (soiled combs and soiled entrance),
　　　　- dead bees outside hive entrance (after cleansing flights).
　　4. Colonies that die out with Nosema succumb due to the viral diseases which take over after

the colony has been weakened due to the original Nosema infection.

• Diagnosis of Nosema can only be confirmed by microscopic examination (for full details reference should be made to our Microscopy Study Notes - the Blue Book).

5.9.2 The effects on the colony.

Because the life of the individual bees is shortened (by as much as 50%) the colony fails to build up in the spring because the worker bees are dying off too rapidly compared with the increase in new brood; the effect is clearly shown in the graph in Appendix 6. Unless samples are taken in the spring it is impossible to determine whether this tardy build up is due to Nosema or due to a scrub queen. A colony with Nosema is not likely to be an effective honey producer during the coming season; however, in our Microscopy Study Notes we have propounded the parameters for light, medium and heavy infections which a competent microscopist should be able to determine and recommend the appropriate treatment.

5.9.3 Treatment of colonies with Nosema.

a) Fumidil 'B' inhibits the spores reproducing in the ventriculus. It does not kill the spores.
b) Autumn treatment: Fumidil 'B', followed by spring treatment the following year.
c) Spring treatment:
    Light and medium infections - Fumidil 'B'
    Heavy -Bailey frame change plus Fumidil 'B'
Fumidil 'B' should be administered in 50:50 sugar syrup. Dosage 166mg of Fumidil 'B' to one gallon per colony.

5.9.4 Other points.

a) Good beekeeping practices prevent the spread of infection in both the hive and the apiary eg. no squashing of bees during manipulations and prevention of robbing, drifting, etc.
b) Disinfection of infected comb and hive parts with 80% acetic acid (100ml/brood box for one week). Note that acetic acid should be placed on top of the frames because the fumes are heavier than air and sink to the bottom between the frames. Acetic acid is very corrosive and metal ends should be removed from frames and any remaining metal work should be greased before treatment.
c) Due to the high incidence of Nosema in UK it is virtually essential to monitor twice a year by taking samples in spring and autumn and treating as required.
d) It is understood that using Fumidil 'B' regularly as a prophylactic for Nosema is unlikely, but not impossible, to produce forms of *Nosema apis* resistant to the antibiotic Fumidil 'B' (Prof. L.Heath).
e) It is important to ensure that any Fumidil 'B' used for this treatment is not time expired and has been stored in a cool dark place.
f) Queens become infected and requeening an infected colony must be regarded as part of the treatment for this adult bee disease.
g) When infection is fully developed after about 10 days from emergence, between 30 and 50 million spores may be found in the ventriculus.

5.10 Describe Acarine, its detection and recommended method(s) of control.

5.10.1 General.

Acarine- causative agent - *Acarapis woodi (*Rennie) - a mite in the class Arachnida.

• These mites (c. 150μm x 65μm requiring a microscope to see them) were discovered by Dr. Rennie at Aberdeen University as a research project funded by the philanthropist Mr. Wood. It is usual for a new biological discovery to be partially named after the scientist who did the research work. In this case both the scientist and the philanthropist are named. This work on Acarine (called at that time 'the Isle of Wight Disease') was commissioned in 1921 after many colonies had been wiped out in UK.

• The EU terminology for Acarine is acariosis and there were proposals, in 1990, that it should become a notifiable disease. At the date of writing nothing more seems to have been heard about this absurd proposal.

• The female mite enters the first thoracic spiracle and then into the associated trachea where it lays its eggs and feeds on the haemolymph of the bee by piercing the trachea walls. The trachea can become completely blocked by the mites and their offspring which can only be seen by dissection and examination under a microscope.

• Signs of the disease: Despite the large number of references to the signs of Acarine in beekeeping literature, Dr. Bailey's work has shown that there are no visible external signs of this disease. The disease has no effect on the flying ability of the bee but it does shorten its life slightly (time not quantified).

• The following signs are those of Chronic Bee Paralysis Virus (CBPV) often confused in much of the classical literature with Acarine:

>   1. Bees crawling with fluttering wings.
>   2. Bees clinging to plants or blades of grass near the hive.
>   3. Bees with a shiny exoskeleton and bloated abdomens.
>   4. Crawling bees outside the entrance to the hive, may be in large numbers.
>   5. Bees with dislocated or partially spread wings (K-wings).
>   6. Bees huddled together on the top bars or on top of the cluster in the hive (these bees do not move away from smoke).

• Diagnosis of Acarine can only be confirmed by dissection and microscopic examination of the first thoracic trachea. When the disease is present the trachea will be discoloured and not the normal creamy colour of healthy adult bees. The trachea can be infested either on one or both sides. For further information candidates should refer to our Microscopy Study Notes (the Blue Book).

• It is important to ensure that samples sent to the microscopist are as fresh as possible in order to make the dissection much easier. Old samples starting to decompose are difficult to dissect.

• It should be noted that there is no correlation between the *Acarapis woodi* and CBPV and *Acarapis woodi* has not been proved as a vector for the spread of CBPV. This is very curious because when crawling bees, etc. are found in a colony and the bees are examined, in a large percentage of the cases Acarine will be present.

5.10.2 The effect of Acarine on the colony.

The effect of Acarine on the colony during the active season appears to be minimal and in a good season with good honey flows can disappear completely. The disease takes its toll if the colony goes into winter and the disease is not treated; something which is unlikely to happen these days when every beekeeper is treating his colonies with Bayvarol/Apistan during August each year.
Conversely, colonies can be clear of Acarine in the spring and infestations can be found in the autumn. This happened to us a few years ago when an apiary of 12 stocks were all clear when samples were taken in the spring but all were found to have Acarine when autumn samples were taken.

5.10.3 The treatment of Acarine.

All the old treatments for Acarine such as Frow Mixture, Folbex (Chlorobenzilate) or Folbex VA (Bromopropylate) have been banned by the EEC on the grounds that they are all carcinogenic thus leaving the United Kingdom without an approved medicament. It is regrettable that the CSL has done no work or given any lead on either Acarine or Nosema over the past decade to assist beekeepers in the UK and the situation presently prevails whereby there is no officially recommended treatment for Acarine which is still an endemic bee disease.

However, our experiments in Devon have shown that the approved treatment for Varroosis, namely Bayvarol or Apistan, is an effective acaricide against this mite. It is possibly because beekeepers are now treating their colonies regularly with these synthetic pyrethroids that little Acarine is found today. We believe that it is also killing off the common bee louse which a decade ago was so much in evidence in most colonies.

5.11 Describe Chalk Brood and Sac Brood, detection and control measures.

5.11.1 Chalk brood.

Known as *Ascosphaera apis* but was originally known as *Pericystis apis*. Most, if not all, of the recent work on chalk brood has been undertaken at Plymouth University under the guidance of Prof.L.Heath.

Causes.

• The presence of spores on nurse bees, combs and hive parts of an infected hive.
• The spores alone are insufficient to cause the disease.
• It is a stress disease (nb. temperature, CO_2 and protein).
• Temperature drop from the normal brood nest $T = 35°C$ to $30°C$ is sufficient for spore germination.
• Spores form in spherical aggregates within dark brown-green spore cysts known as the fruiting bodies which are about 60µm in diameter. The spherical aggregates which contain the spores are c. 12µm in diameter and the actual spores are bacterial size 1.9 to 3.2µm.

• Spore cysts are produced by a sexual process and can only be formed if 2 complementary strains are present (+ and - not male and female) which cannot be identified apart until reproduction starts. Both strains are white but where the mycelium touch a black fruiting body will form.
• Spores are sticky and found all around the hive, on the combs and on the bees. Spores are fed to the young larvae by the nurse bees. These spores germinate given the right stress conditions. The larva dies after the cell has been capped. The fungus grows throughout the larva and the remains are hard and chalky filling the complete cell.
• The cells are uncapped and many of the mummies are removed by the house bees.

Signs:

• Larvae die after the cell is capped.
• An occasional cell is infected or large areas of brood.
• Cappings removed by the bees.
• Larvae become chalky white, fluffy and swell to fill the hexagonal cells.
• Larvae then shrink and harden.
• When infected with two strains of fungus, the colour becomes dark grey or black.
• Larvae removed by the house bees.
• Dead larvae (mummies) are found outside the stock or in a badly infected stock on the floor board.
• Common in the UK especially in the spring or in newly made nuclei.
• Dry discoloured pollen pellets are sometimes confused with 'mummies' but on closer inspection the layers of pollen of different colours are easily discernible.

Treatment: Nil.

Other points:

•Normal brood nest temperatures of 33° to 35°C are likely to be the best prevention to avoid the spread of the disease.
•We believe that it is necessary to recognise two of the stresses necessary for spore germination (CO_2 and temperature) in respect of nuclei. When designing a nucleus box CO_2 is important to ensure that the design reduces it to a minimum and when making up a nucleus temperature is important to ensure enough bees to maintain the brood at the correct temperature. Our own design of nucleus box and care in making up a nucleus has resulted in a much reduced incidence of Chalk Brood fatality.
• Some strains of bee seem to be more resistant than others and a colony susceptible to Chalk Brood will perform very badly compared with colonies of other strains of bee on the same site. As an example see an apiary analysis on page 51 in 'Nudge nudge, hint hint' published by Northern Bee Books.

5.11.2 Sacbrood.

Causes:

• The disease is caused by Sacbrood virus (size c. 30nm) when added to the brood food of young larvae. 2 day old larvae are the most susceptible and the larvae die shortly after the cell is sealed.

• The virus multiplies in the body tissues and appear normal until the cell is sealed when they are unable to shed their last larval skin because the endocuticle remains undissolved. This is when they die.

• Each dead larvae contains about 1mg of Sacbrood virus enough to kill all the larvae in about 1000 colonies.

• Cells containing dead larvae are uncapped and the larvae removed by the house bees. In those larvae remaining, the virus loses its infectivity quickly in the dried remains. It is the adult bees that carry the infection where the virus multiplies without causing distress.

• The youngest workers are the most susceptible when removing dead larvae. The virus collects in the hypopharyngeal glands and is transmitted in the brood food.

• Infected bees cease to eat pollen and quickly cease to feed larvae. This appears to be a survival mechanism at work.

• When they become foragers they fail to forage for pollen. Again the survival mechanism at work; ie. the collected and stored pollen would become infected.

• The lives of infected workers are shortened due to protein deficiencies.

Signs.

• The infection interferes with the moulting process and the final larval moult does not occur (ie. 5th moult after the cell is sealed) with a result the larva does not pupate and dies stretched out in its cell.

• Cells containing dead larvae are uncapped by the bees.

• As the moulting fluid collects between the body and the unshed skin, the colour changes from pearly white to pale yellow.

• After death, a few days later, the colour changes to dark brown; the head changing colour first.

• Finally, the larva dries down to a flattened shape, in the lower angle of the cell, with a slightly upturned head (Chinese slipper effect).

• In the yellow colour stage it can be confused with EFB and in the Chinese slipper stage it can be confused with AFB.

• Disease is very common in UK (30% of colonies are likely to have it).

• It is likely to be noticed more often in the spring and early summer when the ratio of brood to adult bees is high.

Treatment.

There is no known treatment for Sacbrood. Requeening is said to be effective in severe cases.

Other points.

• Sacbrood infection is widespread throughout the world as well as in UK.

• When it appears in a colony it remains generally as a light infection, appears in May and disappears in late summer.

• Drones appear to be unaffected by the virus but large quantities are found in their brains whereas the hypopharyngeal glands in the worker is the main source of infection.

5.11.3 Control measures for both Chalk Brood and Sacbrood.

As there is no treatment the beekeeper must rely on the naturally resistant strains of bee if either of these diseases becomes a problem. Re-queening with another strain is the only option available. As both diseases are endemic, the maintenance of strong stocks and nuclei will avoid colony stress and in the case of Chalk Brood, in our opinion, serious attention must be given to the design of nucleus boxes.

5.12 Describe what action can be taken by beekeepers to avoid damage to honeybees by spraying.

5.12.1 General.

The short answer to this question is that the beekeeper should belong to a spray liaison (or warning) scheme which is the early warning and then closing or moving the colonies to avoid spray damage.

It was probably the increase in acreage of oil seed rape grown in the UK that prompted the formation of spray warning schemes as they are often called. The situation has changed radically since c.1976 when MAFF advised farmers to get to know their local beekeepers and to warn them in advance of any likely spraying operations. This advice was given in 'Approved Substances for Farmers and Growers' which we understand is published annually. By 1983 the advice had changed to get to know your local beekeepers and ask them to nominate a local spray liaison officer. The responsibility changed, over a 10 year period from MAFF to the beekeepers.

Our own personal opinion in this matter is that the onus should be on DEFRA (ex MAFF), and not on the beekeeper, to set up a scheme within their own organisation which is structured on a regional basis and has the resources that are non-existent in the BBKA associations. For example mapping facilities are required together with continuing daily intelligence updates with immediate action on the updates using an efficient communication system. DEFRA have the information and logistics. The BBKA associations do not have them. Additionally, there are about the same number of beekeepers outside the BBKA as those within and details of both these are maintained on a DEFRA database of all beekeepers. However, it is quite clear that the BBKA has taken responsibility for spray warning schemes and the success of these is very varied through the country. We understand that Essex has a well tried and efficient scheme but the one we know about in Devon is a dismal failure.

The only publication on the subject that we have been able to find is 'Spray Liaison' by Malcolm Russell published by the Essex BKA. We are not aware of any BBKA leaflets or directives on how such schemes should be set up and structured.

Most, if not all, spraying is undertaken by spray contractors who are specialist in the application of agro-chemicals. Occasionally farmers or their staff have a certificate of competency and undertake the work themselves.

It is beyond the scope of this syllabus to postulate a spray liaison scheme which are complicated and never seem to work efficiently. The best course of events is to have a good working relationship with the landowner/farmer where your apiary is sited and discuss the problem with him. This is what we do and have never had any real problems and have found it unnecessary to join a spray liaison scheme.

5.12.2 The practical measures possible to prevent damage when prior notification is given of spraying toxic chemicals on crops being worked by honeybees.

There are 5 main aspects whereby the beekeeper can minimise damage to his colonies, these are:

1. Collaboration with farmers within the flying distance of the hives.
2. Participating in the spray liaison scheme if this is operative.
3. Partially closing up the colonies concerned.
4. Completely closing up the colonies concerned.
5. Moving the colonies more than 3 miles while spraying is undertaken.

Collaboration with farmers. We believe that it is incumbent on all beekeepers to make contact with farmers on whose land their bees are likely to forage. They should provide their name, address and telephone number to facilitate the 48 hour warning of spraying which is a statutory requirement on all farmers and sprayers. A pot of honey does not go amiss at such a liaison meeting. In our experience such liaison is appreciated and subsequently the farmers have asked us whether it would be in order for them to spray! All now spray in the late evening or at night as a result and we have never had to close a hive or move one.

Spray liaison scheme. Our own county, Devon, must be rated as one of the failures to make the scheme work and we have little experience with an efficient county scheme. However, having said that we have heard good reports of other counties. Details of the working of these schemes varies from county to county but in essence it requires the liaison officers to maintain a register of hives, beekeepers and sites so that there is a central point of contact for all spraying operations. It is a BBKA scheme and, of course, there are still those outside the scheme who are not members. In Devon, we have now persuaded our RBI to be the Spray Liaison Officer because in his official work he makes contact with beekeepers who are not members of the association.

Partially closing up the colonies. Large quantities of grass/straw are placed in front of the hives covering up and obstructing the entrances. At best it delays the bees going out to forage. We do not consider it a very satisfactory method.

Completely closing the colonies. This is also fraught with danger particularly if the colony is a large one. The basic requirements are as follows:

• Additional comb space must be provided to contain temperature rises.
• A good supply of water must be provided in the hive for cooling purposes (suitable container and sponges to prevent drowning when the water is taken).
• Good ventilation to remove excess heat.

• Entrance to be completely blacked out not allowing light to enter.
• Large quantities of insulation (straw, etc.) over the hives to prevent internal temperature rise due to the sun.
• The absolute maximum that the colonies should be closed in summer is 24 hours.
• All precautions are aimed at keeping the temperature rise to a minimum. Unless this can be contained, there is a very real danger of melted combs and honey and the colony drowning in the sticky mess.

Moving the colonies. We believe that the amount of work involved is less than trying to close the colony up and a much safer approach to the problem. The prudent beekeeper will always have an alternative site arranged for just such an occasion.

See also BBKA Advisory Pamphlet No. 27 'Protecting Honeybees from Pesticides'.

5.13 Describe the signs that suggest a case of poisoning. Describe the actions that should be taken. Describe how a sample of affected bees is collected, packaged and labelled and where this is sent.

5.13.1 Signs of poisoning.

The natural substances collected by the honeybee where poisoning can occur are pollen, nectar, water and man-made sugars that are inadvertently toxic and collected by the honeybee.

The pesticides and herbicides (including fungicides) are a long complex list of chemicals and trade names which have been approved for use by farmers and growers which need not be repeated here or learnt for examination purposes. It is useful to know that the basic technology all started with compounds of arsenic but is now very complex with the sprays based on:

chlorinated hydrocarbons, phosphates, carbamates and pyrethroids.

The death of an insect is, in general, caused by failure of the alimentary system and eventual starvation or by poisons affecting the nervous system leading to complete lack of co-ordination of the normal bodily functions and again leading to death by starvation.

Pesticides and herbicides poisoning.

• Only laboratory test and analysis can provide a satisfactory answer.
• A few or many bees die quickly which can happen suddenly depending on the poison and how much has been taken in to the colony or how many foragers have been affected.
• The number of foragers at the entrance is often less than normal.
• Poisoned bees from inside the hive are ejected.
•The first signs are large numbers of dead bees outside the entrance which usually occurs when the weather is good and the bees are foraging.
• The quantities can be very large (measured in litres of bees).
• Dead bees usually have their proboscis extended.

• The poisoned bees will not be admitted to the hive and are likely to be crawling, trembling, falling over and spinning round on their sides. Can be confused with CBPV.
• The colony is likely to become aggressive and throwing out infected bees inside (fighting).
• If it is only the foragers that have been poisoned the colony will start to recover in c.2 weeks as new brood hatches out and house bees become guards and foragers by natural progression.
• If the house bees have been poisoned, the colony will be very depleted of bees, brood will be uncapped and dead larvae and pupae will be in evidence dead in the cells due to starvation. Finally, there will be only a few bees left with the queen (not usually affected due to her royal jelly diet) and they may abscond.
• Honey is not usually affected. Generally poisoned bees are not admitted into the hive and therefore not unloaded by the house bees. Thus there is no transfer of food to other workers providing an automatic protective mechanism.

Natural poisoning.

• The signs are very similar to those above but generally not as intense.
• Often bees can be found under the offending plant or tree providing the toxic nectar or pollen.

In both cases many bees are lost in the field and poisoned bees do not act normally (Nb. aggression and foraging behaviour).

5.13.2 Action to be taken when spray poisoning is suspected.

1. Comply with the agreed procedure of your local spray warning scheme.
2. Record as much detail as possible about the incident because if litigation is involved it will be some considerable time in the future.
3. Photographs of the colonies and the sprayed crop are often overlooked and are extremely useful at a later date.
4. A large sample is required for reasons outlined above; BBKA Advisory Pamphlet 27 advises 3 samples each of c. 300 bees to be sent to the NBU with the Spray Incident Report which should include the following details:

> Time and date discovered.
> Number of hives affected plus observations on each.
> Estimate of dead bees from each hive.
> Condition of bees and colour of the pollen sample from dead bees.
> Behaviour of colonies (eg. temper, bees being ejected, etc.).
> Sketch map of area and OS grid references showing apiary and crop (don't forget to mark North).
> Weather conditions (wind speed and direction, temperature, rain/fine/sunny/etc.).
> Discuss with crop owner and seek confirmation of spraying and the spray used.
> Visit site with owner, if possible, and determine crop acreage and weeds treated.
> Determine method and time of application together with the flowering state of the crop (nb. photograph of crop).
> Names, addresses and telephone numbers of all concerned including any witnesses.

5. It is important to advise your Branch Secretary and/or your Spray Liaison Scheme representative in order that they may alert other beekeepers in the same area.

6. Don't forget to label the samples and mark the hives!

7. We believe that it is prudent to keep an additional sample in the deep freeze for the Public Analyst in the event of litigation. We understand that samples sent to the NBU will only be used by them in litigation protecting wild life by prosecuting the sprayer or farmer. Any compensation to the beekeeper at law may well depend on this sample kept in reserve and will be the subject of your own proceedings not those of NBU.

8. Feed any poisoned stock which is alive with 50:50 strength syrup immediately, it will dilute any residual toxins in the honey sacs of foraging bees.

9. Maintain hive records of recovery or if and when the colony succumbs.

5.13.3 Collecting the sample, packaging and labelling.

a) There is no special way of collecting a sample as implied in the syllabus. When a real case of poisoning occurs there are literally thousands of dead or dying bees outside the hive. Scoop up 3 lots of 300 to 400 bees and place them in 3 separate paper bags; it is important to ensure that they are not put in plastic bags. Place them in the deep freeze compartment of the refrigerator until they are packed and despatched. Put a label inside each bag and label also the outside of the bags.

b) The 3 paper bags should be enclosed in a suitably sized cardboard box and posted immediately to CSL with a covering letter and full details. It is important that the bees are as cold as possible when they go into the post to ensure that a minimum of decomposition takes place during transit. Recorded delivery is advised for two reasons; firstly, they will be handled expeditiously by the Post Office usually within 24 hours and secondly, there will be no doubt about them being received. Such detail can be important in litigation.

c) Keep another three samples in your refrigerator in case CSL lose the original ones; it happened to us on one occasion and we didn't have a back up!

5.13.4 The expert services available to the beekeeper at national, county and branch level.

At national level the following organisations can be contacted in the event of information being required on spray poisoning and diseases:

> **National Beekeeping Adviser**, CSL, National Bee Unit, Sand Hutton, York YO41 1LZ; for analysis of samples for all diseases and poisoning incidents. It should be noted that charges are levied for most of their services and these charges should be ascertained before entering into a contract with DEFRA (ex MAFF) for any services required.
>
> **IBRA** which is now located in Cardiff; an extensive library and many publications for sale are available.
>
> **BBKA** can provide many useful publications and also provides initial advice on legal matters. Contact should first be made with the General Secretary at Stoneleigh.

At county level, advice and assistance is available from various organisations depending on which county you reside in:

> **County Beekeeping Association** (contact the Secretary).

District or Branch Association (usually the first person to contact would be your own Secretary).

Local DEFRA Offices (to obtain assistance for suspected foul brood infection). It should be noted that most counties only maintain this assistance on a part time basis from April to October. There is no charge levied for suspected foul brood inspections by these officers.

County Bee Instructors (CBI) or Lecturers (CBL) will give advice if they are available. Many of the posts are now abandoned and where they do exist many are part time only.

Agricultural Colleges can often give assistance. Some of the CBIs are based at such a college.

The secretaries of most district and branch associations are in a position to provide addresses and telephone numbers as they receive (or should receive) all the up to date information.

**   **   **   **

HONEY AND HONEY PROCESSING

The Candidate will be able to:

6.1 Demonstrate the apiary equipment normally used specifically for the production of honey, including super comb frames and spacers; section apparatus; queen excluders, devices for clearing bees from supers. Discuss their use.

6.1.1 General.

The main purpose of keeping bees is that a surplus of honey may be removed by the beekeeper for sale or his/her own use. Colonies of honeybees are also kept for scientific, pollination, educational or esoteric purposes. All livestock and humans vary in their productivity and characteristics. The beekeeper needs to be able to recognise the basic requirements of a colony of honeybees in order to:
• Maintain the health of the colony.
• Prevent swarming so that the maximum number of foragers is available at the right time to collect nectar to be processed into honey.
• The beekeeper needs to cull queens which produce bad tempered and swarming progeny. Bad tempered bees make inspections and swarm prevention management virtually impossible. Swarming colonies and/or diseased bees do not produce a surplus honey crop.

6.1.2 The basic requirements for a productive colony of bees are:

• The colony should be 'bubbling over' with bees.
• The colony should be disease free.
• The queen should be young and fertile producing sufficient queen substance to satisfy the needs of the colony and producing progeny with little inclination to swarm, good temper and resistance to disease characteristics.
• The brood box should be sufficiently large so that the queen's laying rate will not be hindered by insufficient comb space.
• The bees should be encouraged to store any surplus stores in the supers. This can be achieved by the use of a dummy board during the rapid expansion in the spring. Adequate space must be allowed for the conversion of nectar to honey. Super early in the spring but do not over-super towards the end of the season. Too many supers will result in the bees storing honey in a column through the centre of the supers.
• A queen excluder should divide the brood box from the supers. This confines the brood to the brood box and reduces the pollen stored in the super frames. Some pollen is usually found in the first super owing to the extension of the curve of pollen from the brood nest below.
• The honey stored in the supers must be free from any contamination eg. sugar syrup, Fumidil 'B,' flumethrin etc.
• The external structure of the hive parts containing the colony should be weather and insect proof except for the entrance.
• The internal surfaces and frames should all be arranged so that 'bee space' is observed in all directions facilitating easy inspection with the minimum of disturbance to the colony and occupying the minimum amount of the beekeeper's time.

• Inspections for swarm prevention and control of swarming should be carried out throughout the swarming season.
• Records should be kept at each inspection noting the presence of the queen, the expansion of the brood nest since the last inspection, correct ratio of eggs, larvae and pupal stages of the brood, any signs of disease in brood or adult bees, any signs of swarming and that there are adequate stores to last until the next inspection.

6.1.3 Apiary equipment normally used specifically for the production of honey.

Ideally each apiary should have a shed where equipment not in use can be stored. This store should be water, insect and rodent proof. All the surplus equipment should be stored in good condition, clean and ready for use. An area should be set aside for the making of frames, sections and repair of equipment. If a separate shed is not available then part of a garage or outbuilding should be set aside for storing all the paraphernalia of beekeeping.

6.1.4 Basic equipment required:-

• Moveable framed hives. All in good repair.
• Disease free bees with current year queens.
• At least four supers available for each stock.
• Spare equipment for making artificial swarms or 'Demaree' method to prevent swarming. This includes spare brood boxes, frames with comb or foundation, floor boards, crown boards, clearer boards, spare queen excluders and roofs.
• Smoker, fuel, hive tool, bee jacket.

6.1.5 The supers.

The supers where the bees store the surplus honey, should have been cleaned after the previous year's extraction ie. propolis and wax removed from wood frames by scraping with a sharp knife, stored to prevent damage or invasion by wasps, ants or wax moths. If any super has been used for the production of brood either accidentally or, deliberately using one and a half brood boxes, these combs will be liable to invasion by pollen mites and moth. See sections 1.9.3.3, 2.17 and 2.20. It is a good policy to use supers only for storage of honey. The production of wax by the honeybee involves the consumption of honey. Supers filled with frames of well drawn out comb are a beekeeper's treasure chest. The spacing of the frames in the supers can be by castellated spacers fixed to the side bars of the super, by metal or plastic spacers or by using Manley frames. These last mentioned frames are self-spacing and avoid the labour of removing and replacing spacers when extracting.

6.1.5.1 Section apparatus.

Producing honey sections requires a strong colony, a good nectar flow in an area free from rape flowers, such as hives placed on the heather moors or good clover country. Types of apparatus used include:-
• frames which can accommodate three sections (hanging section frame)

205

• 'cobana' plastic round section holders
• section crate which can accommodate 28 sections made from basswood.

Very thin wax foundation is used either as a starter or complete foundation in order to avoid spoiling the finished section, for the consumer, with a thick layer of wax in the centre. Cut comb honey is easier to produce. Use thin wax foundation starters in frames placed in the centre of a super.

6.1.6 Queen excluders.

The purpose of a queen excluder is to exclude the queen from the supers while allowing worker bees access to them thereby keeping all the brood rearing and associated pollen in the lower area of the hive ie. the brood chamber. Theoretically only honey would be stored in the supers but in practice it is found that the first super often has quite a few cells with pollen stored in them. There doesn't seem to be any answer to this if the brood nest extends close to the top of the brood frames because the bees will always store pollen directly adjacent to the brood where it is required for use. The queen excluder is attributed to Abbé Collins in France in the year 1865.

The general principle of the queen excluder is a flat sheet with slotted holes just large enough to allow a worker to pass through (it not only prevents the queen passing through but drones as well). Zinc sheet is a popular material; the size of the slots was $^5/_{32}$in (4mm) or 0.156in but most slotted types are now made with slots of 0.162in or 0.163in (4mm) depending on which book is read. The same effect can be obtained with a grill of parallel wires. Note the problem of converting to metric.
• Slotted types: Generally made of zinc but galvanised mild steel slotted excluders are now available. They were made in two versions, one with a series of short slots c. 1½in (38mm), and the other with long slots, c. 3in (76mm). The design is for bottom bee space hives to allow the flat sheet to lay directly on top of the frames. The long slot variety was easily damaged and the short slot version is generally preferred. It is possible to frame this type of excluder as a DIY job; they are not available commercially in a frame. They are the cheapest of all excluders to buy. The mild steel short slot version is probably the best of those available.
• Wire types: These are constructed with strong rigid wires to prevent damage and bending of the wire during use. The construction must be able to withstand damage by burr and brace comb when it is removed from the hive during manipulations. The gaps between the wires must not be greater than 0.165in (4mm). All wire type excluders are framed and should have a bee space on one side only; some are on the market with a bee space on both sides which is wrong and they should be avoided. The framed wire type is known as the Waldron excluder and similar types from Germany are known as the Herzog excluder. Both are more expensive than the slotted types. A further type with wood/wire/wood construction is available in USA at an even greater cost; it is claimed the bees 'like it better'(?) than other types. Better ventilation through the hive is achieved with the wire types compared with the slotted types.

6.1.6.1 Each stock of bees needs a queen excluder. It is best to have an extra couple in the store for manipulating stocks of bees eg. queen rearing.

6.1.6.2. Note care of queen excluders:

• If the excluders are electroplated with zinc (and most are) they should be cleaned with boiling water or the careful use of a small blowlamp. They should not be scraped, the plating will be damaged and rusting will occur.

• When replacing an excluder during a colony manipulation ensure that the top bars are clear of brace comb which may distort the excluder or, worse still, damage it.

• It is quite amazing how often the excluder is put on the wrong way round, the bee space should be below on a bottom bee space hive. Which way should it be placed on the frames; parallel to the frames or at right angles to them? In practice this does not matter as there is a bee space on both sides of the excluder when in use.

• At one branch meeting the Authors saw a Waldron type excluder jammed full of dead worker bees stuck in the slots between the wires. It turned out that the beekeeper had been sold an excluder for *Apis cerana* which is a smaller bee cf. the *Apis mellifera* - most unusual!

6.1.7 General points on clearing bees from supers. See section 2.16

• By definition, clearing implies a crop has been collected and the flow is over; robbing can easily be started unless care is taken when removing the crop.

• Bees are generally more irritable after the flow and will be more inclined to defend their stores than before the flow finished.

• Entrances must be reduced at the same time that supers are being cleared.

• If more than one super has been used it is common for brace comb to have been built joining the supers together, the brace comb being filled with honey. It is essential to remove this brace comb from the top and bottom bars of the frames 24 hours before clearing, in order to avoid honey dripping from the supers as they are removed and starting a bout of robbing. It is a sticky job to do but well worth having the frames cleaned up and no honey dripping while they are being collected. The brace comb should not be there and emphasises the importance of bee space and the many incorrect frames that are in use. The process known as 'cracking the supers' is not given the attention it needs in modern bee literature. It should be done just before dark. It also prevents the surprise of finding the supers not cleared because of brood in the one next to the brood chamber if the first supers are checked during the cracking process.

• Supers should be removed very early in the morning before the colony has started flying and taken straight to the extracting room for extraction the same day.

6.1.7.1 Clearer boards.

There are basically two types, one using Porter Bee Escapes and the other called the Canadian type with long tunnels for the bees to traverse to get from one side of the board to the other.

• Porter bee escapes: possibly the most popular device in UK for clearing bees. The following are the salient points about its use:

a) The phosphor bronze springs require very delicate adjustment to a gap of $^1/_8$in (3mm) and to be free of propolis and wax if they are to work satisfactorily. The vertical alignment should also be checked to ensure that the two springs are central in the case.

b) Two Porter bee escapes per board should be used for rapid clearing and to ensure that if one escape becomes blocked the other one will still be operative.

c) Any clearer board should have an internal bee entrance incorporated in the design with an opening and closing device which can be operated from outside the hive. When wet supers are returned to the hive, the entrance is opened allowing the bees to enter the supers and, conversely, it is closed when they are to be cleared again. It is important that the operating lever allows the roof to be put in place when the supers are off the hive.

d) Approximately 24 to 48 hours are required to clear the supers. The time depends very much on the weather and the flying conditions at the time the board is put on, the better the conditions the shorter the time required to clear.

e) The bee escapes will require cleaning from time to time. Methylated spirit is an ideal solvent for propolis and wax.

• Canadian clearer boards: have the advantage of no moving parts to be propolised and go wrong. The salient points of this mode of clearing are as follows:

a) The same time is required (perhaps marginally shorter) to clear; however, if the weather is bad they are not as effective as the Porter bee escape. The bees seem to learn very quickly that they can return to the supers via the same exit route. The supers must be removed at the latest after 48 hours.

b) An entrance, to and from the brood chamber, capable of being opened and closed from outside the hive is required identical to the board with Porter bee escapes.

c) If by any chance there may be an odd drone in the supers, they can traverse the exit route without blocking it as would happen with a Porter bee escape.

• 8 way plastic escape: which is pinned to the underside of the board directly below a suitable hole. There are 8 plastic slots for the bees to reach the brood chamber and there are again no moving parts. The principle is the same as the Canadian clearer board. Our findings are that it works no better than the Canadian board.

• Scottish clearer board: has 4 holes in each corner leading to a narrowing channel which reduces to a bee space over a distance of about 3in (76mm). We have found that it works no better than the Canadian clearer board despite what the Scots say about it being superior to any other.

6.1.7.2 Shake and brush method.

The method appears to be simple and indeed is, if used at the right time.

a) A spare empty super is required, to receive the cleared frames, placed on a roof behind the hive (note the roof is not upturned as most books recommend) with a cover cloth over to prevent any flying bees re-entering the cleared frames. The colony is smoked first at the entrance and then at the top to drive the bees downwards in the supers. One frame at a time is shaken free of bees and those remaining on the frame are brushed off with a feather. The frame, free of bees, is then placed in the empty super. As one super is cleared so that becomes the receptacle for the next and so on.

b) If the supers are sealed smoking has little effect on the bees; they are only subdued when they have gorged themselves with honey and when smoked they are not immediately subdued, only driven downwards. After the honey flow has ceased the colony is likely to be more aggressive and will defend their stores. It will be clear that it is not a method to be used by the uninitiated at the wrong time and certainly not in an urban situation.

c) Where should all these bees be shaken? We like to shake them back into the hive rather than at the front, only the bees which we brush off land up at the entrance of the hive. The reason for this is that we keep the super covered with a cover cloth except when a frame is being shaken and if the bees are in the hive, we are in control of the situation and not the bees.

d) The final consideration is when should this method be used? At a time when the bees are not flying to safeguard against robbing being started; this means early morning or late evening.

6.1.7.3 Other clearing methods.

To provide an overall picture other methods of clearing should be noted:

a) Mechanical blowers usually powered by electric motor which in turn is powered by a portable generator. The super is stood on end and the bees are blown forcibly out of the super towards the hive entrance. Not for the small time beekeeper.

b) Chemical repellents. The three most commonly known are:

Carbolic acid - not used these days.

Butric anhydride - popular in USA.

Benzaldehyde (smells of oil of bitter almonds) - used quite extensively in UK and works well. Should be kept in the dark. The residual crystals which dry out on the clearer cloth are a fire hazard, therefore the cloth should be kept in a sealed tin.

All the above chemical repellents are used by shaking a few drops onto a cloth which is put over the top super and covered with the crown board. When the bees are cleared (a few minutes normally) the super is removed and the cloth and crown board put over the next super below and so on.

6.2 Discuss the preparation of colonies for specific nectar flows in the area.

This subject has been dealt with in detail in section 4.2

6.3 Demonstrate the equipment used to extract and prepare the honey produced in the apiary and show the place used for processing and packing honey.

6.3.1 The general principles involved.

6.3.1.1 Uncapping the sealed honey frames.

There are a variety of methods for decapping combs, some for the hobbyist and others for large scale operation where a permanent installation is necessary. Some of the more common methods are as follows:

• Various types of knives used either hot or cold and with sharpened or serrated cutting edges. The common feature is the length of blade which requires to be about double the width of the super frame.

• Heating for the knives can range from the simple method of dipping into hot water to electric elements designed into the blade or steam being passed through the blade. The price for this last mentioned piece of equipment is over £100. The electric elements are usually controlled by thermostat.

• The electric carving knife with the two reciprocating blades adjacent to each other works well if cleaned after each cut in hot water. The movement of the blades immediately after cleaning and prior to uncapping throws any remaining drops of water off the blades. We have used this method for many years.

• Decapping machines based on a heated reciprocating blade in a fixed position, the comb being passed over the blade.

• Flailing machines which rip the cappings off are also very efficient for large scale operations. The latest type using rotating nylon brushes costs over £400.

• There are a variety of multi-pronged forks eg. 'Smith scraper' for scraping off the cappings. Useful on badly drawn comb, heather honey combs or combs filled with granulated honey.

• The use of a propane torch passed rapidly over the comb face to melt the cappings has been used in Finland. The cappings form beads around the cell edges. Very little straining of the honey is required after extraction and all the wax is returned to the bees and no receptacle is required for the cappings.

• All other methods require some form of tray or receptacle to receive the cappings.

6.3.1.2 Uncapping trays.

The essential feature of these trays is to provide a receptacle for the cappings to fall into in order to separate the honey from the wax capping. The two basic methods are cold straining through a suitable straining cloth or melting the cappings with the honey and allowing them to separate on cooling.

• The Pratley tray uses the heating principle. It is constructed in stainless steel with a water bath and heating element on the underside. It is expensive and probably one of the worst designed pieces of equipment available. There is no thermostatic control for the heater so one is continually switching on and off. The tray is built on a slope with the thin part of the wedge (and minimum thermal mass) at the place where maximum heat is required, with the result the whole thing clogs up and slows down the extracting process. The separated honey is heated to such an extent it is only good for cooking.

• Most of the devices make no provision for locating the frame over the uncapping tray and it is necessary to fix a bar across with a suitable hole, about $^1/_4$in (6mm) deep, in which to rest the lug of the frame while actually cutting the cappings off.

• An alternative is to hammer a nail through the centre of the wooden bar and use the exposed sharp end on which to pivot the lug of the super frame. Continual use of the spike method will damage the lugs of the frames. Super frames with drawn foundation are a valuable commodity. When well made with glue and non-ferrous metal pins they will last for years. For this reason we prefer the wooden crossbar with central circular depression.

• The beginner to beekeeping should be warned of the pitfalls in this area before he parts with his money; unfortunately most of the experienced beekeepers have learnt the hard way.

• For a beekeeper with 3-5 colonies (9-15 supers) a large stainless steel bowl, 18in (450mm) or more diameter, a sharp bevel-edged bread knife and a wooden cross piece for supporting the frame would be adequate. The honey and cappings which collect in the bowl can be strained through a fine nylon filter.

• The cold straining method seems to be as good as any, the honey is not ruined and the cappings can be washed for mead making or given back to the bees to clean up.

6.3.1.3 The actual operation of removing the cappings.

The important thing is to arrange the set up so that the actual operation is undertaken comfortably with no strain, everything at the right height and everything to hand. It is as well to study the flow of the work from the full supers coming into the extracting room to the empty ones going out.

• Having two people working on the extracting makes life much easier, one decapping and the other operating the extracting machine.

• Most of the books recommend that the frame is held at an angle of about 30 degrees to the vertical and the cut made downwards away from the end of the frame which is being held. The other way is to cut upwards (which we find easier). In both cases the cappings fall away from the frame into the uncapping tray and do not stick to the uncapped comb.

• Again, most of the literature states that the cut should be made just under the capping (in the air space) to minimise the amount of honey removed from the comb. This also is considered to be a matter of preference. We do a straight and level cut across the face of the comb to end up with a set of even combs in the super ready for next year's operations. There are always likely to be some uneven combs and this is a good time to put these to rights.

• Manley type frames help to provide a straight cut, the uncapping knife being guided by the wooden bottom bars and the top bar of the frame.

• If the knife is heated in hot water it is as well to have a clean cloth close by to wipe the knife dry before making a cut, continual drops of water will increase the water content of the honey. This is not necessary with the reciprocating blades of an electric carving knife.

6.3.1.4 Separating the honey from the cappings.

Wax cappings should be separated from the honey and rendered down carefully using the minimum amount of heat. The wax can be moulded and used for showing; it is unadulterated with other hive products and requires little cleaning. High temperatures change the colour of the wax, deepening the shade. Wax for show should be pale, clean and with a delicate aroma. Iron containers should not be used.

• The cappings can be cleaned up by the bees in say a Miller type feeder with a grill to allow the bees to access both sides. They may require a stir after a couple of days to let the bees get to the inner sticky portions.

• Cappings can be left to strain overnight through a fine filter into a settling tank. The almost honey-free cappings can then be washed in cold water. Then separate the water and wax using a fine nylon strainer. The honey water so obtained can be used for making mead. The cappings when dried can be set aside for show wax, making candles, cosmetics or polish.

• Other methods of separation included using centrifugal force and spinning in a special cage in the extractor, using a purpose built decapping tank with inner mesh for straining the honey away or using a converted spin dryer or washing machine as recommended by DIY enthusiasts.

6.3.1.5 Other points.

• Extracting is a sticky, thankless task and care should be taken to keep everything scrupulously clean and tidy.

• Store the supers awaiting extraction in a warm dry room. Warm honey is easier to extract.

• Any honey which has been heated in the Pratley tray should only be used or sold as cooking honey (baker's honey).

• Rape honey which has been allowed to granulate in the combs cannot be extracted by the spinning method. Heating the entire comb in a stainless steel tray, not unlike a jumbo size Pratley, will separate the honey from the wax, no uncapping required. Controlled heating of the comb to allow the wax to melt at about 147°F (63°C) followed by rapid cooling of the honey in order to prevent the level of HMF increasing and reduction of diastase activity requires specialised equipment. In order to avoid any unnecessary heating of the honey, the crop from the rape fields should be removed and extracted as soon as the honey is 'ripe'. Strain and store the honey, before it crystallises in air-tight containers until it is required for bottling.

6.3.2 Honey extractors.

6.3.2.1 General principles.

The principle involves two forces which occur when an object is rotated eg. a ball on a string. The first is the centrifugal force which acts away from the centre of rotation and the second is the centripetal force which acts towards the centre of rotation. In the case of the ball, the centrifugal and the centripetal forces act on the ball and because they are equal the ball stays where it is on the end of the string. In the case of a honey extractor, only the centrifugal force acts on the honey and it moves outwards away from the centre, the droplets of honey hit the drum wall and trickle down to the bottom of the extractor under another force, gravity. The centripetal force, of course, is acting in the rotating frame of the extractor which being rigid does not move. Honey is hygroscopic. Extraction of honey should be undertaken in a warm dry atmosphere. Large commercial honey packers in USA or Australia use honey drying stores for supers awaiting extraction to prevent any increase in the water content of honey.

6.3.2.2 Types of extractor available.

6.3.2.2.1 Tangential extractors.

Generally have a cage whereby the frames for extracting can be inserted at a tangent to the circle of motion allowing only the honey on the outside face to be removed during operation of the extractor. The honey on the inner face is pushed on to the comb septum during operation thus necessitating the frame to be reversed in order to complete the extraction. The salient points of this type are as follows:

• With full frames the first side can only be partially extracted at slow speed otherwise the weight of honey on the inner face is likely to break the comb.
• The combs must be reversed to extract the second side and then reversed again to complete the extraction of the first side. This is time consuming.
• It is essential to have wired combs for use with this type of extractor.
• The number of frames that can be extracted is limited by the size of the cage, 6 being the realistic limit.

6.3.2.2.2 Radial extractors.

These extractors are all very similar with a rotating framework designed to hold the frames radially with their top bars vertical and parallel to the sides of the drum and their side bars on a radius of the extractor's rotating framework. The salient points are:

• Honey on both sides of the comb is extracted simultaneously and there is no reversing of frames which makes the process that much quicker.
• Combs that are unwired can be easily extracted without damage.
• The design permits a greater number of frames for a given diameter drum compared with a tangential, 10 being a realistic size for a small model (or one super).
• Speed control is not as critical as the tangential but nevertheless it is necessary to start slowly.
• For a given speed and number of rotations, the radial is inferior on the amount of honey remaining in the comb.
• Unwired combs can be extracted by this method but care must be taken with new comb which is less strong and may be broken as a result of dynamic unbalance. This unbalance occurs when pollen or granulated honey is present in the combs. Extra care is necessary when increasing the speed to prevent damage to the combs. Only liquid honey can be removed by spinning.

6.3.2.2.3 Brinsea horizontal extractor.

Spins two frames horizontally in a plastic bowl with a honey gate. Works well for a small producer. Many small time producers like to produce sections or cut comb to avoid the necessity of extracting the liquid honey.

6.3.2.2.4 Other points of interest.

• Materials for construction range from tin plate, galvanised steel and polythene to stainless steel. These days only the two latter materials are used. Stainless steel costs more but is likely to prove the most economical in the long run.
• Both types can be either hand or motor driven. If the electric motor is used a speed control device is necessary to give a range of speed from 0 to about 400rpm. The problem arises in getting a device to give high torque at low speed when it is starting up from rest. The horse power [hp] required from the motor is quite small, a fraction of a hp being adequate.
• When purchasing a radial type, care should be taken that it will accommodate the frame the beekeeper intends to use. Many (most?) are designed for $^7/_8$in side bars and the full complement of Manley frames (10 super frames) cannot be loaded because of their wide side bars ($1^5/_8$in) .
• Most tangential types can take a reduced number of brood frames which is often useful; only some of the radial types can do this.
• Both types require to be loaded giving the best dynamic balance. No matter how carefully the frames are selected and loaded it will never be perfect and the extractor at top speed wants to move. There are two solutions, one to screw it down to a platform with castors and let it move or two, to bolt it securely to the floor. Our own is fixed in three places with rigging screws and chain from the top rim of the extractor.
• If a fixed installation is preferred then it is desirable to raise the extractor so that a honey bucket can be put straight under the outlet tap without any lifting.
• Never leave any rape honey in the bottom reservoir of the extractor. Strain and store in buckets as soon as possible.

• Never leave the extractor unguarded when the honey tap is open or the frames are being spun.
• Most enterprising beekeeping associations have an extractor for hire to its members for a small fee.

6.3.3 Extraction of Heather Honey.

Heather, *Calluna vulgaris,* or ling honey is thixotropic (property of becoming temporarily liquid when shaken or stirred and returning to gel state on standing). When bottled after extraction its colour, aroma and typical aerated composition distinguish it from floral honeys. Because of its gelatinous nature ling honey cannot be extracted in the same way as floral honeys. The combs of heather honey can be processed in the following ways:
• Cut the sealed comb from the frame and place in a strong straining bag and press the honey out. For small quantities of honey the press can be made from two wooden planks hinged at the wide end or for larger quantities a wooden 'Peebles' type press (cost c.£400) which will press 3/4 combs at one time.
• Scraped back to the septum with a 'Smith' cutter. The honey and wax placed in a strong straining bag and spun in a metal cage in the extractor,
• After uncapping the cells of honey can be agitated by a 'Perforextractor'. This is a roller, shaped like a rolling pin, with fine steel needles attached to the barrel in a pattern to match the honey comb. When the roller is passed over the uncapped comb of heather honey the needles enter the cells and agitate the honey. The action of the needles liquifies the heather honey allowing the comb to be extracted in a tangential extractor.
• Sold as cut comb honey.

6.3.3.1 Other points to note:

• Because of the higher water content (up to 23%) of heather honey there is a greater danger of the honey fermenting.
• Supers should be stored in a warm dry room prior to extraction.
• Ling heather honey makes very good cut comb and used to be sold in Scotland cut directly from the comb.
• Use a strong fine mesh straining bag when pressing the comb and collect the honey in clean containers in order to dispense with the necessity of any further straining.
• Brother Adam's hydraulic press at Buckfast Abbey in South Devon will press 2 tons of honey per day with a loss of no more than 1.2% of the crop.
• Honey from the bell heather, *Erica cinerea* is not thixotropic and can be extracted as for floral honeys.
• The honey collected by the bees moved to the moor to collect heather honey may not be 100% ling heather. If the heather is not yielding then the bees will look for other forage.

6.3.4. The straining and settling of honey after extraction.

This removes solid matter down to a particle size determined by the strainer. There are three types of solids, those that sink to the bottom, those that float and those that remain in suspension. The solids can include wax, bees, grubs, propolis, sugar crystals, wood chips and other extraneous matter. All extracted honey must be strained and allowed to settle. The best time for this is immediately after extraction. Different strainers will remove different solids. Coarse and fine stainless steel strainers are usually used for the initial straining before storage in honey buckets. Before bottling, 'run' honey should be warmed sufficiently to pass easily through a fine nylon strainer, 80 mesh to inch, to remove any small particles of dust, wax, crystals of sugar etc. If these small particles are left in the honey the honey will remain cloudy and lack sparkle. The presence of the particles will encourage crystallisation. The particles act as a nucleus onto which the crystals of sugar will granulate and grow.

6.3.4.1 Methods of straining.

•. Single settling tank. The tap is at the bottom of the tank and only those solids that float will be strained out including air bubbles. The honey passes through a fine straining cloth as it is poured into the tank. Allowing 24 hours in a warm room for settling of the newly extracted honey is usually sufficient to remove the air bubbles.
• Sump tank with baffles. The input compartment is usually at a lower level than the outlet. The tank contains 3 or 4 baffles giving 4 or 5 compartments. The baffle openings are alternately top and bottom with the honey flowing alternately under and over the baffles. The surface can be skimmed as required to remove floating debris while the dense solids collect in the bottom of the tank. The tank can be double walled to provide a water jacket for heating if required. This method is used for large scale production of honey.
• Filtering under pressure. Diatomaceous earth filters are used by commercial honey packers. The honey passes through the filters at a high temperature and is then immediately and rapidly cooled. Most of the pollen as well as the extraneous matter is removed by this method resulting in a very clear end product which has lost much of its flavour and colour.
• Wire and cloth strainers. These can be in a variety of formats depending on the scale of the operation. Wire strainers should be made of stainless steel or monel metal. Cloth filters are usually made of nylon. The old fashioned use of 'cheese cloth' would leave small fibres in the strained honey. Commercial strainers such as the one designed by Ontario Agriculture College in Canada (O.A.C. honey strainer) incorporates a series of concentric metal strainers of different mesh size. The honey flowing through the coarse mesh first and the finest mesh last eg. 12 mesh/inch to about 80 mesh/inch.
• 'Strainaway'. Made by Garn Products. This is a fast method of straining honey by creating a vacuum in the honey collecting chamber. Figures quoted for straining 30kg of honey at 86°F (30° C) are 4 minutes compared to 120 minutes using a settling tank and a nylon bag filter. This equipment is light and easy to handle making it easier to clean. It is made from food grade polythene and stainless steel. The filters range from a basic 1,600 holes per square inch to an extra fine filter with 10,000 holes per square inch.
• The most simple type. Small time beekeepers strain the honey as it goes into the settling tank. It has one major drawback; it produces a large number of bubbles which are formed when the honey drops from the straining cloth into the settling tank. If a long sausage shaped filter is used then the sealed end can be keep below the surface of the honey and air is not introduced during the filtering process.

6.3.4.2 Other points in relation to straining are:

• The higher the temperature the easier it is to strain with a fine mesh. A temperature of 95-100°F (35°-38°C) is considered to be satisfactory. The viscosity of the honey increases rapidly with temperatures below 90°F (32°C).

• Well filtered honey will take longer to granulate.

• If a honey such as rape is to be stored in buckets and bottled later it is better to complete the fine straining before it is stored ie. while it is still in its liquid state. Later it only needs warming to a point where it will flow for bottling, this stage is reached well before all the crystals have melted. In this state it would be impossible to pass it through a strainer without bringing it back to the completely liquid state and thereby heating it unnecessarily. Using this method the honey is sold as 'soft set' honey.

• Long straining cloths can be an advantage. When clogged the cloth can be pulled across the tank to an unused portion. All straining is best below the surface of the honey to prevent bubbles of air forming.

• Large commercial honey packing organisations use very fine filters by pumping the honey under pressure at high temperatures and then cooling it quickly after straining. Such methods are not a practical proposition for small scale operators. This liquid honey has a wine-like clarity which is very attractive to the buyer.

• Always avoid adding air to the honey during processing. Do not allow the honey to fall from a great height. Use an elongated filter bag which remains under the surface of the honey.

6.3.5 Extracting room.

The need for good hygiene in the handling of honey for human consumption is really common sense and can be defined under the two following headings, namely:

1. To ensure that the beekeeper does not infect the final packaged product with foreign bodies of any kind, particularly those organisms appertaining to diseases of mankind.
2. To ensure that the beekeeper does not run foul of the law and be subsequently prosecuted under United Kingdom or European Economic Community (now European Union) legislation which is generally administered and policed by environmental health officers and personnel.

Both the above are inter-related by virtue of the fact that the legislation is drafted to avoid problems of foreign bodies in a natural food.

Most beekeepers undertake the processing in the family kitchen. It prompts the question whether it is suitable? Are there two sinks (one for washing utensils and the other for washing hands) and are surfaces of the walls and floors washable? It is likely to be taboo if there is a washing machine installed where soiled linen is washed despite the fact that all the family food is prepared every day in the same room. Beekeepers must be vigilant of current legislation, it is all too easy to transgress unwittingly. In the future, EU regulations may require all beekeepers producing, processing and selling honey to be registered.

6.3.5.1 The common sense approach.

Tools, utensils, equipment and packaging. This is a straightforward washing up job with lots of hot water and clean cloths. No soap or detergent should be used to ensure that no contamination can occur with cleansing agents of any kind. The packaging of the honey in glass jars and plastic cartons is the real danger area as most beekeepers assume that they are clean when received from the supplier. In many cases they are bought in bulk from the manufacturers and boxed up into smaller quantities by the bee appliance suppliers. The boxing up process often introduces foreign bodies and dirt. Beware of small chips of glass in honey jars and cracked jars; they should all be turned over to remove any chips and then washed and polished before filling. Even when the jars arrive sealed in polythene the same care should be given to the jars before use. Lids will also require inspection. A good deal of work can be eliminated if the jars and lids are stored in a clean dry area before use. The extractor, filters, settling tanks and storage containers should be cleaned of all honey immediately after use, dried and stored carefully away until required.

Hands and clothing. It is understood that most people catch a cold or flu or other viral infection by infecting themselves with their hands which in turn have picked up the infection from another source. Consider the common cold; it would be an unusual individual who did not infect their hands by using a handkerchief or tissue. These infected hands then use the handle of a door covering it in large quantities of 'bugs'. Along comes the next person to use the door handle and pick up the 'bugs' on their hands. They infect themselves by touching their face, mouth, eyes, etc. Shaking hands with somebody who has a cold is a fine way of catching it from them. A grand old merry-go-round. Wash your hands and then don't touch any part of your face. If you do, then wash your hands again before continuing with the task of processing honey. Use a clean overall especially for the job; regulations these days would require all operators, male or female, to wear a suitable protective hat. Smoking is, of course, a no go activity.

Clean area. The work area, which for many hobbyist beekeepers will be the home kitchen; very few will have a discrete extracting and honey processing room custom built for the purpose. Do ensure that everything is spotless before operations start, paying particular attention to all working surfaces. All pets should be banished. Windows and doors, if not screened, should be kept closed during extraction and bottling of honey. There should be no trace of vermin, flies, bees, wasps, ants, spiders etc. either in the extracting room or in the room where the honey is stored. No other activity should be taking place in the room where the honey is being processed.

Personal hygiene. Any one selling honey to the public may have their premises inspected by the Environmental Health Officer. Personal hygiene is paramount, dishevelled hair, dirty unkempt nails and hands (broken skin), stained clothing, signs of smoking, obnoxious body odours or signs of ill health eg. coughing and sneezing, should not be tolerated by anyone assisting or operating the extracting and bottling process. There should be a separate toilet close by the extracting or bottling room with hot and cold running water where the operators hands can be washed after each visit.

6.3.5.2 It is very important to keep up to date with the regulations which are constantly changing to 'harmonise' (a popular bit of Euro-talk) them throughout the Community. The BBKA publicise these changes in BBKA News and attention is brought to the County Association Secretaries and to Branch Secretaries. If you are in any doubt, ask your Branch Secretary. See also section 6.8 below.

6.4 Have available for inspection by the Assessor, typical samples of packed honey ready for the table and for retail sale.

6.4.1 The candidate shall provide samples of his/her honey bottled and ready for sale.

The examiner will be looking for any faults in the samples that are presented for examination. As there is plenty of time to prepare the samples before examination day, no candidate should fail in this part of the syllabus. Note that only bottled honey is required and this alone must be produced for examination.

The amount is specified as 6 jars minimum. We would suggest two jars of liquid honey, two jars of set and 2 jars of creamed honey would, in our opinion, satisfy the requirement for this part of the syllabus.

If it is prepared with the attention that one would give the honey if it was being entered into a show there should be no problems. It must be free from any contamination with extraneous solids (good filtering requirement) and be legally labelled with undamaged lids having clean wads. There should be no incipient granulation in the liquid honey. Note that the 'flowed in' plastic seal lids without wads are acceptable.

Although the syllabus does not specifically state that the candidate is required to discuss the samples, the candidate should be prepared for this. Typical questions could be as follows:

> What type of honey is it? Identification by pollen analysis is not required and an inspired guess usually fits the bill if the source is unknown.
> How was it strained and what size mesh was used?
> How old is it and how was it heated?
> How is the honey sold? At the door and/or through a retail outlet?
> What are the regulations for selling at your own door?
> How is the weight checked?
> Etc.

6.4.2 Discuss with the examiner the types of honey obtained in the candidate's district.

We would expect a further discussion about the samples presented in respect of the type of honey collected in the district concerned. The district concerned is the area described by a 3 mile radius from the apiary or within flying distance from the hives.

It is always a good idea to examine an ordnance survey map (preferably 1:25000) and then examine the area on foot to see what plants, trees and shrubs are available for forage. A photocopy of the area concerned with any major forage sources marked on it would convince and, no doubt, earn some brownie points from the examiner.

The colour of the honey should be known, light, medium or dark and whether there is any honeydew collected in the area. The names of the crops that the bees work should be known together with the colour of the associated pollen loads.

6.5 Describe the arrangements made by the Candidate for extracting honey from the comb and the preparation of Comb Honey.

6.5.1 Extracting honey from the comb.

This topic has been fully discussed in section 6.3 above. The Candidate will be required not only to describe how he/she extracts honey from the comb but will normally be required to show the Examiner where it is normally extracted.

6.5.2 The preparation of section, cut-comb and chunk honey for sale.

6.5.2.1 Preparation of sections.

• The sections must be fully sealed (capped in wax by the bees). If open half-filled cells are present, the section is not suitable for sale. Honey is hygroscopic and these cells will absorb water.
• The woodwork should have been treated with paraffin wax before the sections were placed in the hive. This will protect the woodwork leaving it unstained when the propolis and paraffin wax is scraped off.
• Each section should be held up to the light and any with cells of pollen should be rejected.
• All sections should be put into the freezer (minimum 10° C) for a period of 24 to 48 hours to ensure that any *Braula coeca* and wax moth eggs are killed. If this is not done, it is possible for the section to be ruined while on the shelf awaiting sale.
• Pack the sections into 'Tupperware' type containers to avoid damage to the surface of the comb whilst it is in the freezer. During the thawing process (on removal from freezer) first allow the container to reach room temperature before opening and removing the sections. This prevents water condensing on the surface of the comb.
• Before packing and after cleaning, the section should be weighed and the net weight of honey comb noted for labelling.
• The packing should be in specially made cardboard boxes with a built in cellophane window on one side to display the sealed comb. These boxes are obtainable from the equipment supplier.
• After packing into the container it must be labelled correctly in accordance with the regulations.
• The special cardboard containers usually describe the product adequately and the only addition required is the name and address of the producer (telephone number may be included) and the net weight.

6.5.2.2 Preparation of cut comb.

• The combs of honey from the supers are selected for thickness, good cappings and freedom from pollen and granulated honey. Any combs that do not meet these criteria should be used only for extraction.
• A flat board or 'Formica' surface larger than a super frame is required for preparing the cut comb. The tools required are a Price cutter (the size to suit the clear plastic container, nb. two sizes of container are now available) or a large sharp knife and a template as a guide for cutting the comb and a kitchen spatula for moving it after cutting.

• The comb in the frame is inspected and placed in the middle of the board with the best side upwards. The knife is run around the inside woodwork of the frame cutting the whole comb free from the frame which then simply lifts off. This can be returned to the bees for cleaning up with the rest of the wet supers.

• The comb is then cut into pieces to exactly fit the containers. A piece of comb approx 4 in×3 in×1^1/$_2$in is needed to fill a 227g (8oz) cartoon. This is easily done with the Price cutter which not only cuts it to size but provides a tool to place it accurately in the container all in one operation. If a Price cutter is not available, the comb can be cut by knife and placed in the container by using the spatula. A fair degree of expertise is required to not only cut it to the right size but then to get it into the container without damage to the comb. Before the cut comb is put into the container, it should be drained for a few minutes to allow the honey to run off the cut edges. Using the Price cutter does not allow this to be done easily. There should not be a line of honey in the bottom of the container after it is filled with comb.

• The surface of the comb should be immaculate and free from drips of honey and any damage.

• The packed cut comb should be frozen for a couple of days to kill off any *'Braula coeca'* present.

• The label placed on the outside after any condensation has evaporated.

• The labelling requirements are the same as for sections. Cardboard sleeves are now available, but add to the cost of the product. The minimum description must be 'COMB HONEY' followed by net weight. A black on white 'Ablelabel' with a space to put the actual weight in by hand is acceptable. A second label with the producer's name and address and lot number is necessary. There must be more illegal labelling in cut comb than in any other honey product on the market.

• There is a certain amount of wastage from one BS shallow frame when used for cut comb; the surplus can usefully be used as chunks for 'chunk honey' and it is advisable to have some 454g(1lb) jars handy for the chunks, the jars to be filled later with honey when extraction has been completed.

• To prevent granulation if the comb honey has to be stored for long periods, it should be kept in the freezer at low temperatures. Honey granulates fastest at 57°F(14° C). Above and below this temperature granulation is slower. Higher temperatures are not good because of the ageing effects (diastase and HMF) on the honey. Low temperature storage is the best.

• Heather honey is an ideal honey for cut comb because it is thixotropic (in a jelly form until stirred when it will become liquid and flow for a short time reverting back to jelly) making it easy to cut up with no waste of the actual honey.

• Rape honey granulates rapidly in the comb. It is better to extract this honey while it is still in the run state. Cut comb from the rape blossom does not sell well. Honey show judges prefer the exhibits of comb honey which contain liquid honey.

• All comb honey should be produced in wire free super frames which have been fitted with extra thin foundation or 'starters'.

• Drone or worker foundation can be used. When pulling out drone comb the bees use less wax than pulling out worker foundation. The disadvantage of drone comb is that should the laying queen find her way into the supers she will fill the supers with drone brood.

6.5.2.3 The preparation of chunk honey for sale.

• One piece of sealed comb approx. 1in× 1^1/$_2$in× 3in for the traditional squat (454g) honey jar. Straight sided jars with a wide neck are sometimes used making it easier to introduce the piece of comb honey.

•The comb must be surrounded with a heat treated liquid honey otherwise granulation takes place and the product becomes unsightly.
• Fill with heat treated liquid honey. The liquid honey needs to be well filtered and 'bright' to show off the comb in the jar.
• Hold the jar close to the honey tap and run the liquid honey down the side of the jar to avoid introducing air bubbles.
• The major fault is to provide too small a piece of comb. Do not use comb honey containing pollen.
• The total weight should be carefully checked. The honey comb tends to rise above the level of the liquid honey.
• The labelling must observe all current regulations.
• Prepare small quantities of chunk honey and supply it to the retailers in small quantities. Chunk honey has a short shelf life, the run honey tends to granulate and the appearance of the comb is spoilt.

6.6 Describe the processing and storage arrangements for the honey and packaging for sale.

6.6.1 The processing of honey.

This is addressed in sections 6.9.1 and 6.9.2 below and 6.5 above.

6.6.2 The storage of honey.

6.6.2.1 Principles of storage of honey is to prevent any deterioration taking place ie:

• Any increase in the HMF factor. This is a result of ageing and heating of honey.
• Any decrease in the diastase activity of the honey. This is also a result of heating and long term storage.
• Any fermentation of the honey. This is due to a water content higher than 20%, the crystallization process of honey and the presence of yeasts.
• Any contamination of the honey due to storage in faulty or inappropriate containers eg. used previously for storing strong smelling foodstuffs eg. vinegar.
• Any loss of the aroma and taste of honey due to non air-tight containers.
• Any destruction of comb honey due to the presence of wax moth or *Braula coeca.*

6.6.2.2 Storage of extracted honey.

Honey should be stored in 30lb white polythene (food only) buckets with well-fitting lids. Anything bigger becomes difficult to handle when preparing the honey prior to bottling. The old method used 28lb tins with lever lids with attendant problems of the lacquer and plating becoming faulty and the subsequent rusting. Full honey buckets should be stored in a clean, cold below 50°F (10° C), vermin proof store until required for bottling. Each bucket should be marked with date of extraction, source and type of honey.

6.6.2.3 Storage of comb honey.

Comb honey can be stored in the freezer. This will kill off the wax moth and *Braula coeca* and keep the honey in a liquid state. First place the comb in airtight plastic bags or 'Tupperware' type boxes. This will protect the surface of the comb during the storage at low temperatures and when removed for use. Thus water condenses from the warm moist air outside the freezer onto the container not onto the comb. Honey is at its best when first taken from the hive.

6.6.2.3 Other points are:

• The buckets require to be as full as possible to minimise the amount of air trapped in the top.
• Before the lid is snapped shut, the centre of the lid should be depressed onto the honey to minimise the air content.
• Store at a temperature of 57ºF (14ºC) for rapid granulation and then at as low a temperature as possible after it has set. Don't open to check the granulation, do it by feeling the sides of the buckets. When honey is in the run state the sides are quite flexible but very solid when granulated.
• Long term storage should be below 50ºF (10ºC) to slow down any tendency to ferment and to minimise ageing of the honey (HMF and diastase content).
• Store honey for a maximum of 12 months. As honey ages the HMF value increases and the diastase activity decreases.
• Never extract unripened honey or allow the water content of honey to exceed 20% by careless handling and storage. Honey is hygroscopic and if left exposed will absorb moisture from the atmosphere. A refractometer will give an accurate reading of the moisture content of honey.
• Sugar tolerant (osmophilic) yeasts are present in all honeys. In honey with a moisture content greater than 20% the yeasts will multiply, secrete enzymes which ferment the sugars in honey creating alcohol, acetic acid and carbon dioxide. This chemical reaction will cause the storage buckets to expand and rupture giving off a foul smell.
• Glucose is less soluble in water than the other honey sugars. Glucose dominant honey granulates faster than honeys with a high fructose content. During the crystallisation process molecules of water are released which increase the water content of the uncrystallized honey. If the honey already has a higher than average water content then any increase in temperature will encourage fermentation. Only extract and store ripe honey.
• Unripe honey should be given back to the bees before it ferments or pasteurised to kill off all the yeasts and then sold as 'Baker's Honey'.
•When preparing stored honey for bottling keep the heating process to a minimum.

6.6.3 The packaging of honey.

The packaging of honey means bottling extracted honey in all its forms and inserting comb honey into cut comb containers and additionally into attractive cartons for sale. These items are addressed in sections 6.9.1, 6.9.2, 6.5.2 and 6.8.

6.7 Describe how the requirements for public health and safety, consumer protection, food hygiene, as overseen by the Environmental Health Officer, apply to Candidates in the area.

6.7.1 General.

This is a new topic in the BBKA series of examinations and one cannot but wonder why it has been included, particularly in an examination which is of a practical and oral nature. Any knowledge of how the EHOs operate in a particular area is unlikely to improve the beekeeping husbandry of the Candidate. If EHOs are included in the syllabus then there seems to be a serious omission and that is in respect of the local Trading Standards Officers who will be concerned with the labelling and selling of honey if there is any transgression of, for example, the Weights and Measures Act.

The primary task of the EHOs with respect to beekeepers is to ensure the enforcement of the applicable legislation as contained in the statutory regulations; those in respect of the preparation and sale of honey are listed in section 6.8.2 below.

We understand that the modus operandi of EHOs may be expected to be the same all over the country no matter in which area the Candidate may reside or keep his bees. In general, they play a passive role and react to complaint situations from the public. Any Candidate for the examination should make contact with the EHO in his area to check whether there is any divergence from our notes. EHOs are employed by the local council, the telephone number will be found in the directory.

6.7.2 Public health & safety.

Public health & safety is primarily concerned with nuisance caused by the actual beekeeper's bees and swarms whether they be in public places or on private premises. Nuisance can be caused by keeping too many stocks adjacent to neighbours and thereby creating situations whereby neighbours and their domestic animals are being stung or washing on the line is being fouled.

Swarms in public places can be a danger to the public and as such fall under the remit of the EHO. Swarms that establish themselves, particularly in chimneys and cavity walls, can also be a danger to the householders concerned.

In general, the EHOs react to complaints received in the matter of public health and safety and initially endeavour to enlist the services of a local beekeeper before calling in a pest control contractor to destroy the new found nest. The EHOs are likely to maintain a list of beekeepers who are willing to assist and always endeavour to save the bees rather than destroy them as they are well aware that honeybees are beneficial insects and play an important role in our food chain as well as the food chain of other animals.

6.7.3 Consumer protection & food hygiene.

Consumer protection and food hygiene are primarily concerned with extracting and packaging of honey for sale. Honey is regarded as a low risk food due to its high sugar content, antibacterial properties and the well known inhibine effect. In the event of a complaint, which usually manifests itself in the adulteration of the product with foreign bodies, the EHO will investigate the beekeeper concerned to identify and correct the transgression.

Only commercial beekeepers who are extracting honey in premises more than 5 days in any five consecutive weeks are required to formally register their premises with their Council. There are exceptions to the requirement to register (see later). However, all premises from which commercial activities are carried out, even domestic premises, are subject to the requirements of the Food Safety Act 1990 and hygiene regulations made thereafter. Such premises will be eligible for inspection. Inspection frequencies are risk based and it is therefore unlikely that regular visits will be carried out due to the low risk honey production poses. Clearly, if premises or production shortcomings are found, visits will be more frequent. EHOs are normally more than happy to give goodwill advice so as to assist commercial producers to meet the various legal requirements.

6.8 Demonstrate familiarity with current regulations and any other statutory requirements as they affect those offering honey for sale.

6.8.1 General.

Statutory regulations regarding the sale of packs of honey are made by governing bodies to protect the consumer/customer from unscrupulous dealers. Beekeepers with surplus honey to sell should make every effort to be well informed on all the current legislation. This legislation appears to be constantly changing especially with metrication and other EU regulations. Satisfied customers will continue to purchase their favourite brand of honey from the same source ie. a producer of a good reliable product. This reputation, built up over many years can be easily lost by carelessness or ignorance. Bad labelling, sticky jars, poor packaging or semi-crystallised honey are not attractive to customers.

Adherence to all the legal requirements regarding the sale of honey will avoid prosecution by Trading Standards or Environmental Health Officers. A high standard of production, prepared with 'due diligence' will protect the beekeeper from a civil prosecution should any damage to the health of the consumers arise as a result of contaminated products. Honey produced by a hobbyist beekeeper and sold at the door should be prepared with the same care as that produced for the retail market. A look at the supermarket shelves with the attractive products of imported and home produced honey at very attractive prices should make the small time beekeeper realise that if he wants to sell his home produced honey at a premium price the presentation of his products must have equal or better characteristics.

From time to time regulations change. The 'BBKA News', produced 4/5 times a year and distributed to all members of the BBKA, reports on new regulations produced by DEFRA (ex MAFF) or other regulatory bodies pertaining to the sale or production of honey for sale. Statutory Orders are obtainable from HMSO or can be viewed at public reference libraries. Copies of the Statutory Instruments are expensive and not easily obtained. At the reference library consult Sweet and Maxwell's Practical Food Law Manual, Butterworth's Law of Food and Drugs, Halsbury Statutory Instruments and the Statutory Instruments (see numbers below). If the beekeeper is in any doubt about his obligations he should seek advice from the Local Authority Officers, LACOTS (Local Authorities Co-ordinating Body on Food and Trading Standards) responsible for supervising the regulations on sale of food. Obtain a copy of BBKA's Advisory Leaflet No 103,

'So you wish to sell honey?' which has been produced with the aid of Devon, Somerset and Warwickshire Trading Standards departments, Devon Beekeepers' Association and Mr. Derek Brown of Reading.

6.8.2 The statutory laws and regulations applying to the sale and supply of honey products:

- **The Honey Regulations 1976.** Statutory Instruments 1976 No. 1832.
- **The Food Safety Act 1990**.
- **The Food Labelling Regulations 1996.** Statutory Instrument 1996/1499 as amended.
- **Food Premises (Registration) Regulations 1991.** S.I. 1991/2825 as amended.
- **Materials and Articles in Contact with Food Regulations Act 1987**. S.I. 1987/1523.
- **The Weights and Measures Act 1985. Primary regulations Weights and Measures (Miscellaneous Food) Order 1988** Statutory Instruments 1988 No.2040 as amended by S.I. 1990 No. 1550. Local working standard weights and measures testing equipment regulations 1986 S.I. 1986 No 1685.
- **Weights and Measures (Packaged Goods) Regulation 1986** Statutory Instrument 1986 No 2049.
- **Weights and Measures (Quantity marking and Abbreviations of Units) Regulation 1987**. Statutory Instrument. 1987 No.1538.
- **Weights and Measures (Metrication) (Miscellaneous goods) (amendment) Order 1994**. Statutory Instrument 1994 No 2866.
- **Trade Descriptions Act 1968.**
- **The Consumer Protection Act 1987.**
- **Lead in Food Regulations 1997 SI 1997/1254 as amended.**
- **Flavourings in Food Regulations 1992 SI 1992/1971.**
- **Food Additives Labelling Regulations 1992 SI 1992/1978 as amended.**
- **Food Safety (General Food Hygiene) Regulations 1995 SI 1995/1763 as amended.**
- **Food (Lot Marking) Regulations 1996 SI 1996/1502.**
- **Plastic Materials and Articles in Contact with Food Regulations 1998/1376.**

Where a date and number for the Statutory Instruments has not been given it is because we have not been able to trace the necessary document at our reference library. The regulations are extremely complicated. The main requirements are explained in the following sections:

6.8.3. The main requirements of the current statutory regulations affecting the preparation of honey for sale.

The Food Safety Act 1990.

This act makes it an offence to sell honey which does not meet 'food safety requirements' by it being rendered injurious, unfit for human consumption or contaminated, making it unreasonable to expect consumption in that state. There is also an offence to sell food which is not of the nature, substance or quality demanded by the purchaser. The most common use of this section relates to the sale of food containing foreign matter. In respect of honey it could be used where the description of the 'honey' was applied to products such as 'honeydew' and polymerised corn oil. It could be an offence to falsely declare the country of origin of honey as in the customer's view the quality of the honey is related to its country of origin. Section 21 of this act introduces a new defence which has not previously been available to food producers, namely the 'due diligence' defence. The effect of this provision is that if any producer in selling a product commits an offence he will not be liable for prosecution if it can be shown that all reasonable precautions were taken and all 'due diligence' was exercised to avoid the committing of the offence. Any product must not be harvested where unacceptable amounts of potential harmful substances are likely to be transferred to food.

The Food Premises (Registration) Regulations 1991 as amended.

These regulations require beekeepers who sell honey to register their premises with the Local Authority ie. premises where the honey is extracted, bottled, stored or sold. New businesses should register 28 days before trading. Amendments exempting hobbyist beekeepers as 'low risk' premises are as published in BBKA News, number 83 February 1992. These exemptions are quoted below:

> "**Premises used irregularly or only occasionally.** Premises used for less than five days in any five consecutive weeks, the five days do not have to be consecutive and thus any premises used regularly once a week will be included in the exemptions.
> **Premises where only low risk activities take place, which are exempt unless the retail sale of food takes place there.** This category includes places where fish is taken for food but not processed, where eggs are produced or packed and where honey is harvested.
> **Domestic premises.** Used for the production of honey or subsequent preparation, storage, bottling or sale (whether wholesale or retail) of honey."

Beekeepers should as far as possible conform to the requirements of the existing regulations concerning the preparation of food as mentioned previously. Local Authority Inspectors are entitled to inspect a beekeeper's premises where honey is being processed. The purpose of these inspections are:

• To identify potential hazards and assess the risks to public health arising from activities within the food business.
• To assess the effectiveness of management control to achieve safe food.
• To identify specific contraventions of food hygiene law, these are namely the requirements of the Food Safety (General Food Hygiene) Regulations 1995.

If a beekeeper or trader applics to be registered the following basic requirements must be met (BBKA News No 77 November 1990):
a) All surfaces must be washable, they must be easy to clean and disinfect, this includes walls, floor and shelving i.e. made of stainless steel or tiles. It must not be possible to crack these surfaces. The ceilings must be constructed to prevent accumulation of dirt and to reduce condensation of steam.
b) Windows and doorways should be screened to prevent entry of insects. Doors must have a smooth, non-absorbent surface. Doors should close automatically and be hermetically sealed.

The premises should be free from vermin.

c) Two sinks with hot and cold water should be available, one for utensils and one for operators. There should be separate toilet facilities.

d) All equipment should be maintained to a high standard of cleanliness. All equipment must be impervious, resistant to corrosion and capable of repeated cleaning.

c) The operator should observe a high standard of personal hygiene. Clean overalls and hair cover are essential. Hands should be washed before commencing work. Nails should be kept short and clean. No cuts or abrasions present on the hands.

d) Where food is being prepared the area should be a 'no smoking area'. There should be no dust, smoke, no spitting or any unpleasant odour which may contaminate the product.

e) Good storage i.e. vermin free, exclusion of domestic animals and weather proof for supers awaiting extraction and storage of packed honey.

6.8.4 The main requirements of the current statutory regulations affecting the labelling and weight of packs of honey. **The Food Labelling Regulations 1996 and the Food (Lot Marking) Regulations 1996.**

The presentation of food shall not be such that a purchaser is likely to be misled to a material degree as to the nature, substance or quality of the food. The important points for labels are:

• The name of the product to be displayed. This may be:

a) Honey   b) Comb Honey   c) Chunk Honey   d) Baker's Honey or Industrial Honey.
e) The word 'honey' with a regional, topographical or territorial reference eg. Devon Honey, Honey from South Devon, Moorland Honey etc.
f) The word 'honey' with a reference to the blossom or plant origin eg. Heather Honey, Lime Honey.
g) The word 'honey' with any other true description eg. Honeydew, Pressed Honey, Set Honey.

• With wholesale transactions of containers of a net weight of 10kg or more, a separate document showing the required information is sufficient if it accompanies the container.
• The containers of honey for sale should be given a 'lot' mark. In the event of contamination all honeys of the same 'lot' number will be recalled from the retail outlets. The 'lot' mark should be easily visible, legible and indelible. The size should be appropriate to the operational pattern. The mark should be prefixed by 'L'. Records should be kept of the lot marks of all honey sold.
• There must be no misrepresentation in words or pictures.
• The producer's name and address to be displayed (telephone number optional).
• The net weight to be displayed in metric units. The size of the lettering for the weights is critical.
• The numerical value may be declared in words, in which case the unit of weight must be referred to in words.
• The numerical value may be declared in figures which in the case the unit of weight may be referred to in words or by means of an approved abbreviation.
• One type space between the figure and the unit used is not mandatory and presentation with the unit directly adjacent to the number may be used.
• All required markings should be clear legible, conspicuous and indelible.

• Honey weighing less than 50g, chunk honey and comb honey may be packed in any quantity.
• The units which may be used for declaring quantities are kilogram and gram. The permitted abbreviations are kg for kilogram and g for gram (lower case only). No other abbreviation should be used. The letter 's' may **not** be added to indicate the plural. Weight markings are subject to a minimum size requirement where the size is that of the number and not the unit as shown below:

Contents of pack	Minimum height of numerals
Not exceeding 50 gram	2 millimetres
Exceeding 50 gram but not exceeding 200 gram	3 millimetres
Exceeding 200 gram but not exceeding 1 kilogram	4 millimetres
Exceeding 1 kilogram	6 millimetres

Rules governing the declarations of quantity were changed on 1st October 1995 by a directive from the European Union (EU). This is the date when the Imperial weights were replaced by the metric system. The net weight of honey in the container should be 57g, 113g, 227g, 340g, 454g, 680g, or a multiple of 454g. The net weight of the honey in the container should comply with the following criteria:

Net weight of product	Maximum scale interval on weighing equipment
50g to not exceeding 200g	2g
200g to not exceeding 1kg	5g
1kg to not exceeding 25kg	10g

Labelling.

Labelling is very much a personal choice, we believe in simplicity and use Ablelabel (32mm x 63mm) black lettering on gold for all our honey. Once you have a label 'stick' with it if your honey sells well. The public quickly learn to recognise the label as a sign of a good product. Where a list of ingredients must be affixed to the product eg. Honey Mustard, Honey Fudge, Honey Linctus then the list must be preceded by an appropriate heading which consists of or includes the word 'ingredients' The listed items should be in descending order of weight.

Sales of home produced honey.

Sale of honey may take place at the beekeeper's door providing the goods for sale have been produced on the premises. It is therefore necessary for the beekeeper to have his hives in the garden or grounds of his home otherwise he will be contravening the retail licensing laws. If all your honey is sold to a retailer then the apiaries may be sited anywhere. The labelling requirements are identical in both cases.

Weights and Measures Act 1985.

Honey is a prescribed food and its packing is governed by the 'average weight' rules. This means that any production batch of honey must consist of packs, the average content of which is equal to the declared quantity. If these requirements are met then the packer is entitled to apply 'e' mark alongside the declaration of quantity. The advantage of this is that anyone exporting to another European country can do so with the knowledge that the product will have free access to the market without further checks.

Weighing of products for sale.

A small time seller of honey will not wish to purchase expensive scales as used by the retail trade because of the cost. Prices range from £200 to £500. Nevertheless the scales used for weighing products for sale should be regularly checked using stamped brass weights. We think for a small producer if the scales were checked twice a year and logged this action would demonstrate 'due diligence'. A commercial firm processing honey for the retail trade would be required to use scales which would meet the approval of the Weights and Measures Department inspectors who would mark the scales with a government stamp ie. crown and date if approved or a six sided star if rejected.

Containers.

Containers for honey should be made of materials which under normal and foreseeable conditions of use do not transfer their constituents to the honey in quantities which could endanger human health, bring about a deterioration in its aroma, taste, texture or colour or bring about an unacceptable change in its nature, substance or quality. This applies to containers which are in contact with the honey and to the containers which are likely at some later time to be in contact with the honey. Certain ceramic materials may present particular risks. Packers are asked to obtain an assurance from their suppliers that containers comply with **The Materials and Articles in Contact with Food Regulations 1987, Plastic Materials** and **Articles in Contact with Food Regulations Act 1992 and The Glazed Ceramic Ware (Safety) Regulations 1975.** With ceramic containers difficulty arises with amount of honey a ceramic container will hold, whether the internal glazing is adequate and the efficiency of the lid to exclude the atmosphere in order to keep the honey in pristine condition until consumed. We have recently purchased pottery containers which would only hold 411g (14½oz).

6.8.5 The main requirements of the current statutory regulations affecting the composition of honey for sale.

The Honey Regulations 1976 seek to regulate the compositional quality of honey. These regulations cover 10 pages and are available from HMSO, Statutory Instrument 1832. The following are summaries from the regulations.

Regulation 2 contains a number of definitions:

• 'Honey' means the fluid, viscous or crystallised food which is produced by honeybees from the nectar of blossoms, or from secretions of, or found on, living parts of plants other than blossoms, which honeybees collect, transform, complete with substances of their own and store and leave to mature in honeycombs.

• 'Comb Honey' means honey stored by honeybees in the cells of freshly built broodless combs and intended to be sold in sealed whole combs or in parts of such combs.

• 'Chunk Honey' means honey which contains at least one piece of comb honey.

• 'Blossom Honey' means honey produced wholly or mainly from the nectar of blossoms.

• 'Honeydew honey' means honey, the colour of which is light brown, greenish brown, black or any intermediate colour, produced wholly or mainly from secretions of or found on living parts of plants other than blossoms.

• 'Drained honey' means honey obtained by draining uncapped broodless honeycombs.

• 'Extracted honey' means honey obtained by centrifuging uncapped broodless honeycombs.

• 'Pressed honey' means honey obtained by pressing broodless honeycombs with or without the application of moderate heat.

Regulation 3 is an exemption clause for honey sold outside the UK or sold under contract to HM government for consumption by any of HM services.

Regulation 4 controls the composition of honey and in particular prohibits the addition to honey sold of any substance other than honey. It also prohibits the sale of honey unless it is as far as practicable free from mould, insects, insect debris, brood or any other organic or inorganic substance foreign to the composition of honey. It also prohibits the marketing of honey which has an artificially changed acidity.

Regulation 5 concerns the use of honey used as an ingredient in the preparation of food.

Regulation 6 deals specifically with the use of the description 'honey'. In simple terms it states that a product may only be called honey if it is honey as defined in regulation 2.

Regulation 7 lays down certain physical characteristics for honey and specifies the circumstances under which the descriptions comb, chunk, baker's and industrial must be used. In the case of baker's or industrial honey it lays down compositional criteria relating to moisture content, foreign tastes and odours, fermentation, reduced enzyme levels, diastase activity levels and hydroxymethylfurfuraldehyde (HMF) content. Note should be made of the standards set:

• Heather honey being derived wholly or mainly from the genus *calluna* or clover honey derived wholly or mainly from the genus *trifolium* may have a moisture content of no higher than 23 per centum.

• Any other honey must have a moisture content of no higher than 21 per centum.

• The diastase activity of the honey should not be less than 4, or if it has a naturally low enzyme content, less than 3.(*This may occur in Hungarian Acacia honey, Robinia and may be due to either the age of the foraging bee or the opulence of the flow of nectar. Enzyme activity is regarded as an indicator of careful processing and storage of honey. The diastase activity of honey can be measured by spectrophotometric test or using a more simple apparatus using degraded starch and iodine).*

• The hydroxymethylfurfuraldehyde (HMF) content should not be more than 80 milligrams per kilogram. A trace of HMF is always naturally present in honey though with adverse and prolonged storage or overheating the quantity can exceed 30-40mg/kg and rise even higher. *(Measured by Fiehe test or Winkler's direct calorimetric test; see the book on 'Honey' by Dr E. Crane, reprinted 1979).*

Regulation 8 prohibits the use of a label which makes reference, either in words or by pictorial device, to the blossom or plant origin of the honey unless the honey is derived wholly or mainly from the blossom or plant indicated. The same prohibition applies where there is a reference to the regional, topographical or territorial origin of the honey unless it originated wholly in the appropriate place.

Regulation 9 provides that any statement required to be marked on a label must be clear, legible and indelible. It must be in a conspicuous position on the label in such a manner that it will be readily discernible and easily read by an intending purchaser or consumer under normal condition of purchase and use. The statement must not be interrupted by any other material where this might mislead the purchaser as to the nature of the honey, neither must it be hidden, obscured or reduced inconspicuousness by any other matter on the label. The height of the letters in any statement must not be such that, by giving undue prominence to any part of the statement will result in the purchaser being misled.

Regulation 10 lays down the penalties for contravening the regulations. The penalties range from a fine not exceeding £100 or 6 month imprisonment to a fine of £500 and a term of imprisonment not exceeding one year.

These Honey Regulations are somewhat daunting to the hobbyist selling a few pounds of honey in the local shop which are surplus to his family's requirements. The bees can produce the product in accordance with the regulations and the beekeeper should only extract it when completely capped (water content-OK), sell it without heating (diastase and HMF-OK) and show 'due diligence' straining and bottling (foreign bodies-OK). However, for examination purposes a knowledge of the regulations is necessary.

Schedule 1 gives a method of determining diastase activity.

Schedule 2 gives the compositional requirements for honey relating to apparent reducing sugar content, moisture content, apparent sucrose content, water insoluble solids content and ash content. Summarised here:

- There should be no addition of substances other than honey.
- The honey should be as far as practicable, be free from mould, insect debris, brood and any other organic or inorganic substance foreign to the composition of honey. Honey with these defects should not be used as an ingredient of any other food.
- The acidity should not be artificially changed. There is a legal maximum level of acidity ie.'not more than 40 milli-equivalents acid per kilogram'.
- Any honeydew honey or blend of any honeydew honey with blossom honey should have an apparent reducing sugar (invert sugar) content of not less than 60% and an apparent sucrose

231

content of not more than 10%. Other honeys should have an apparent reducing sugar content of not less than 65% and an apparent sucrose content not more than 5%.
- Honey with a moisture content of more than 25% should not be supplied.
- The maximum water insoluble solid content is:
  - for pressed honey   0.5%
  - for other honey   0.1%
- The maximum ash content is:
  - for honeydew honey and blends containing honeydew honey   1.0%
  - for other honey   0.6%

Baker's or Industrial Honey.

Honeys of the following descriptions should be labelled or documented only as 'baker's honey' or 'industrial honey':
- Heather honey or clover honey with a moisture content of more than 23%.
- Other honey with a moisture content of more than 21%.
- Honey with any foreign taste or odour.
- Honey which has begun to ferment or effervesce.
- Honey which has been heated to such an extent that its natural enzymes have been destroyed or made inactive.
- Honey with a diastase activity of less than 4 or, if it has a naturally low enzyme content, less than 3.
- Honey with an hydroxymethylfurfuraldehyde (HMF) content of more than 80mg per kg.

The Trade Descriptions Act 1968.

This act imposes a general prohibition on the sale in the course of trade or business under a description which is false or misleading. The falseness can relate to any physical characteristic of the goods such as quantity, size, composition or method of production. It is also an offence to give a false indication as to the place of production or false claim as to the county or country of origin.

The Trade Description Act 1972.

This act has been repealed. This act required that any imported product which bore a UK name or mark had also to be marked with an indication of the country of origin. This means that imported honey blended/ and bottled in UK no longer has to declare the origin of the honey.

Organic honey.

The term 'Organic Honey' is misleading as the range of the honey bee when foraging cannot be strictly regulated. The use of this term would involve considerable risk and should be avoided.

6.8.6 Enforcement.

Inspections are carried out by the local Environmental Health and Trading Standards Departments.

These agencies would respond to any complaint from a member of the public. Government regulations are constantly changing and being updated. It is therefore important that beekeepers make themselves familiar with up to date information before processing, packing and selling honey for retail sale. Operate within the law; litigation is always expensive. Should any customer complain to the producer beekeeper about the quality or quantity of honey or be dissatisfied in any way, the wisest action to take is to return the customer's money or replace the offending product or both. Always seek advice from either the County Secretary or General Secretary of the BBKA should there be any threat of prosecution.

6.9 Describe Liquid Honey and Set Honey (both granulated and soft set) and a method that may be employed to obtain these with good quality results, including mention of the recommended temperatures for satisfactory results.

6.9.1 The preparation of liquid honey.

All 'run' honey should be heat treated before sale to delay granulation for 6 to 12 months. The honey needs to be raised to a temperature of 130°F (54°C) for about 45 minutes. If a higher temperature is used 160-180°F (71-82°C) there is a danger of the honey being burnt, the enzyme activity being destroyed and the HMF factor increase. Partially granulated run honey looks terrible and is a very common fault with the hobbyist beekeeper. There is much local honey on sale which should really be taken off the shelf and returned to the beekeeper as being unfit for sale. Many of these jars are labelled with a County Association label and thereby bring other beekeepers a bad reputation. We believe that a County organisation should take some responsibility for quality if their label is to be used. It is for this reason we don't use our County label.

6.9.1.1 Method of bottling liquid honey.

• Honey being an unstable solution of sugars will granulate after extraction. Depending on the origin of the honey this may be with a small or large crystal. Fructose dominant honeys are slow to granulate and produce a large crystal. When preparing liquid honey for sale it is these fructose dominant honeys which are warmed, strained and bottled.
• Starting with fully crystallised 28/30lb buckets from store, remove from the surface of the honey in the storage bucket all the debris which has floated to the top using a small sharp knife. Clean and replace the airtight lid.
• Warm the containers of honey for 2-3 days in a thermostatically controlled heater at 90°F (32°C). The heater should have a fan to minimise any hot spots occurring.
• An old refrigerator makes an ideal warming cabinet when fitted with a suitable thermostat, a small fan and a couple of 25watt electric light bulb. Our refrigerator doubles up as an incubator for queen cells in the summer by resetting the thermostat. An 'Eco Micro' thermostat with its own heater is more expensive, sophisticated and reliable.
• Check the storage buckets after 24 hours by pressing the outside of the storage drum to assess the length of time necessary to convert the honey back to liquid state. Always use the minimum of heat ie. don't forget you have placed the buckets of honey in the heater.
• Prepare containers to receive the honey. Most hobbyist beekeepers continue to use the squat glass jars. These can be purchased 'film' wrapped. Nevertheless the jars should be inspected for

blemishes and washed in clean hot water to remove any dust. Here a dish washer is ideal as the jars are heat dried and warm for bottling. All lids should be checked that no dust is adhering to the under surface.

• Never use damaged glass containers.

• The bottling should take place in a warm, dust free, odourless room. Keep windows and doors closed to prevent entry of any insect or dust.

• The run honey from the buckets should be strained through a fine filter into a settling tank This will remove any fine crystals or specks of dust which will form a nucleus/focus for granulation if left in the honey. Allow the honey in the settling tank to stand in a warm bottling room for 12-24 hours to allow any air bubbles to come to the surface.

• When bottling, keep the empty jar close to the tap and allow the honey to run down the side of the warm jar.

• Cover the filled jar of honey with a lid immediately to prevent any dust being attracted to the surface of the honey.

• Make random checks for weight with an occasional jar of honey using calibrated scales, weighing the jar empty and after it is filled.

• Remember that when honey is warm the density is reduced and the volume increased. The jars should be well filled above the neck to allow for the increase in density as the honey cools. The test of holding the capped jars to the light to see if any gap is present is not fool-proof especially if the honey is warm or contains air bubbles.

• At the end of a bottling session there is always a certain amount of 'scum' at the bottom of the tank. This is due to the presence of air bubbles. This honey should be drained out of the tank and used for home consumption.

• The 227g (½lb) hexagonal jars on sale from the bee suppliers are more expensive than the conventional one pound squat jars but are attractive to the customer. These jars can easily hold 256g (9oz) of honey. When filling the hexagonal jars air bubbles become trapped in the six upper corners. Use a sterilised curved probe of non ferrous metal to remove these air bubbles before closing with a clean lid.

• Lids should come clean and undamaged from the suppliers. We always make a habit of checking our deliveries and return any damaged lids for exchange. The lids should be checked for dust or foreign particles and applied as soon as possible to the filled honey jar. Tighten the lids after the honey has cooled to make a good airtight seal.

• The 'flowed in' lids make a better seal than the old type lid plus 'waxed wad'. The latter lid still has its uses for honey show exhibits or selling chunk honey to prevent the run honey coming into contact with the lid.

• After bottling, heat treatment is required to prevent granulation for a further six months. Before labelling, the jars complete with lids are placed in a water bath for about 45 minutes. The water should be kept at 130ºF (54ºC). The level of the water should reach the neck of the jars. Commercial processors rapidly heat the honey to 160ºF (71ºC), pressure filter and rapidly cool the honey before bottling.

• Unfortunately the BBKA has not undertaken to give advice on the use of microwave ovens as a method of heating honey in order to delay granulation. Perhaps in the future we may see some authorised work done on this process where guidelines are given in order to maintain the diastase activity and prevent any increase in the HMF content.

• Fermenting honey due to poor handling eg. insufficiently air-tight containers or the honey having been extracted in an unripe state, can be heated to a temperature of 200°F (94°C) for a short time to kill the yeasts but this will not reduce the water content sufficiently to make it completely safe from re-infection by other wild yeasts. It can then only be sold as 'bakers' or cooking honey or used to make mead.

• Beekeepers with only one or two hives will probably wish to bottle their 100 or more pounds of honey directly from the settling tank soon after extraction is finished ie. without storing the honey in 30lb storage buckets. Many beekeepers feel that heating the honey destroys its fine aroma and taste. They are probably right. But if the honey is to be sold heating the jars of honey is the only way to delay granulation taking place for about 6 months. Smoky or partly granulated honey becomes unattractive to the customer.

• Dust particles, air bubbles or pollen grains left in honey will encourage crystallisation.

• Once bottled in air-tight containers the honey should remain in a warm atmosphere for 24 hours to allow any residual air bubbles in the honey to come to the surface, then stored in a dark cool area at about 50°F (10°C).

• When supplying retail outlets ensure that:

    1. The honey stored on the shelves is not exposed to sunlight.

    2. Run honey once bottled has a shelf life of about 3 to 6 months. Supply your local outlets with reasonable quantities of honey at frequent intervals depending on the demand. Avoid long term storage in jars. Always exchange any partly granulated 'run' honey.

    3. Granulated honey should only be delivered to outlets after complete granulation has taken place. Any sudden drop in temperature in partially granulated honey will cause unsightly frosting around the neck and shoulders of the jar. The honey has shrunk away from the glass.

    4. Avoid the use of the labels excusing the granulation/frosting as a natural phenomenon of honey.

    5. A high quality unstable product like honey needs careful handling and storage.

6.9.1.2 Bottling Ling Heather honey.

Much of the heather honey obtained by beekeepers will be sold as comb honey. Being thixotropic it is ideal as cut comb as after cutting the honey does not drain away from the cells in the comb after the initial draining. Heather honey commands a higher price compared with floral honey eg. Bottled in one pound jars £3.50 opposed to £2.50. Pure heather honey has a distinctive flavour described by Herrod-Hempsall as 'bitter sweet'; it is light to dark amber in colour with a pungent aroma of heather flowers. When bottled, glistening air bubbles are present as a result of the pressing process of extraction. Some show judges regard the bigger the air bubbles the better the exhibit. Pure ling honey does not granulate but, when it is mixed with floral honey partial granulation will eventually take place, star-shaped granules appear suspended in a jelly-like medium. Ling honey has a higher protein and water content than floral honeys. If ling honey is over heated during processing much of the aroma will be lost and the thixotropic quality will be destroyed. Honey from bell heather honey, *Erica cinerea* is not thixotropic and can be extracted in the same way as floral honeys. The honey is a port wine colour with a distinctive flavour. The bell heather grows on the heathlands alongside the ling heather; it blooms in June, earlier than the ling.

	Ling honey	**Floral honey**
Protein	1.8 -2 %	0.2%
Water (sealed in comb)	Up to 24 %	17 - 20 %

6.9.1.3 Special care to be taken when bottling ling honey.

• After extraction ling honey should go straight into the settling tank and bottled whilst still warm.
• The honey in the tank should be stirred ready for bottling.
• The lower density of the honey due to the presence of air bubbles increases the volume of the honey. The honey jars should be well filled at the neck and weight carefully checked.
• Because of the higher water content ling honey is liable to ferment.

6.9.1.4 General points on bottling honey for sale.

• When bottling has been completed the outside of the jars should be free from stickiness.
• The lid should be well screwed down making an air-tight seal.
• Suitable labels should be attached on each jar, many labels are too big for 227g ($^1/_2$lb) jars.
• Tamper proof labels enhance the look of the product but increase the cost of production. The design could include the 'lot' number.
• At some time in the future EU regulations may require all beekeepers producing, processing and selling honey to be registered. Most beekeepers undertake the processing in the family kitchen. It prompts the question whether it is suitable? Are there two sinks ? Are the walls and floors washable surfaces? It is likely to be taboo if there is a washing machine installed where soiled linen is washed despite the fact that all the family food is prepared every day in the same room. Beekeepers must be vigilant of current legislation, it is all too easy to transgress unwittingly.
• Prevent partial granulation of run honey by careful preparation.
• Prevent fermentation by storing honey in airtight containers at a temperature below 50ºF (10ºC). Honey with bubbles on the surface indicate fermentation. At an advanced stage fermenting honey has a characteristic unpleasant smell and the surface of the honey heaves!
• Do not use scratched lids and unclean wads.
• Always keep the surface of the honey free from impurities.
• Never sell underweight jars of honey.
• Always ensure the labelling is correct.

6.9.2 The preparation and bottling of naturally granulated, soft set and seeded honey.

6.9.2.1 General.

Some types of honey will granulate more quickly than other types. Fast granulating honeys produce a fine grain crystal, slow granulating honey produce a larger crystal which is gritty to the tongue. Glucose is much less soluble in water than fructose. If there is a higher percentage of glucose than fructose in the honey then there will be fast granulation eg. rape, *Brassica napus* (immediately after extraction). If rape honey is left on the hive too long the honey will granulate in the comb causing further complications.

6.9.2.2 Naturally granulated honey.

Types of honey that are likely to granulate quickly with a fine texture should be strained immediately after extraction. If naturally granulated honey is required then the honey should be bottled directly from the settling tank. This will produce a 'hard set' honey. Sometimes the surface is difficult to penetrate with the honey spoon. Pure rape honey will have a lard-like colour. After bottling the jars of honey should then be stored at 57°F (14°C) to allow granulation to take place. If moved too soon to a cold store frosting will occur. This is a particularly difficult fault to avoid with honey that granulates rapidly and is associated with low temperatures. It is caused by shrinkage, generally at the neck of the jar, as the honey granulates. Honey that granulates slowly or granulates at higher temperatures, greater than 57°F (14°C), seldom exhibit this fault. During the granulation process water molecules are released as the sugar crystallises increasing the total water content of the remaining honey. This may cause fermentation to take place. A vertical streak on the side of a jar of granulated honey with a rough surface texture will indicate that fermentation is taking place. Always avoid scratched lids and unclean wads, underweight products and incorrect labelling.

6.9.2.3 Soft Set Honey.

• To obtain a soft set honey the extracted honey should be fine strained immediately after extraction. Then stored at 57°F (14°C) until granulation has taken place. Then stored at the coolest temperature possible below 50°F (10°C).
• When preparing to bottle 'soft set' honey, warm the buckets of granulated honey in the warming cabinet for 24/36 hours or until the honey attains a pouring consistency.
• For 'soft set' honey, only use honey which has set with a fine grain. This can be determined when cleaning off the surface layer before heating the buckets of crystallised honey. Honey containing rough or large crystals will need to be seeded before bottling for 'soft set' honey.
• The clean warm honey when at a pouring consistency can be poured into a settling tank. Stirred thoroughly and allowed to settle for a few hours in a warm room. The stirring breaks up the formation of the sugar crystals in the honey giving a smooth creamy texture. This honey is sometimes referred to as 'creamed' honey.
• The creamer or stirrer must be kept below the surface of the honey to prevent introducing any air bubbles.
• Electrical heating coils can be used to wrap around the tank to keep the honey warm allowing the air bubbles to come to the surface.
• Bottle after careful preparation of jars.
• The last pound or so of honey in the settling tank may be filled with air bubble sometimes called 'scum'. The last jars may not contain the correct amount of honey owing to the presence of air bubbles. Underweight jars should not be sold.
• This honey which has granulated in the storage bucket and then rewarmed for bottling and stirred, will not return to its previous rock-like state. It is a 'soft set' or 'creamed' honey.
• Be sure that the honey jar lids are tightened after the honey has cooled in the jars.

6.9.2.4 Seeded Honey.

This honey should be of a fine texture. Honey should be seeded with 10% rape or similar honey which granulates with a fine grain eg. clover or dandelion. With so many fields of rape planted every year unless the beekeeper makes a special effort to extract the spring honey before the rape fields come into bloom much of the spring honey will be seeded as it is extracted. Rape honey is ideal for seeding, producing a very attractive granulated honey with fine crystals. If the beekeeper is not in an area where rape is grown then in order to produce a granulated honey which is not too hard, does not contain large crystals, is quick to granulate with an overall consistency and produces an attractive product, he will need to seed his honey. Honey was first treated in this way by a Canadian called Dyce in 1931. Much of the Canadian honey had been lost due to fermentation. He found that honey was less likely to ferment if it granulated rapidly with a fine crystal.

6.9.2.5 Points to observe.

• The honey to be seeded needs to be returned to its liquid state by using the warming cabinet, 90°F (32°C) and filtering to remove any dust particles or crystals of sugar. This honey is allowed to cool in the settling tank.
• Honey is damaged if heated above 95°F (35° C).
• The crystals in the 'seed' the honey must not be dissolved. This honey should be in a soft state.
• Both the honeys should have been strained at the time of extraction ie. when in the liquid state.
• If 45lb (20kg) of honey is placed into a settling tank then 5lb (2.3kg) of the 'seed' honey is added and stirred carefully and thoroughly.
• Avoid adding any air to the blend of honey by keeping the mixer below the surface of the honey.
• Allow the honey to settle in the bottling tank in a warm room for 2-3 hours.
• Prepare all the jars with care as previously stated.
• Bottle the honey while it is still in a warm state.
• Once bottled keep the jars of honey in a warm room for 24 hours to allow any air bubbles to come to the surface. Then move to a temperature of 57°F (14°C) to allow granulation to take place. This will keep frosting to the minimum.
• Once granulation is complete the honey can be stored at a low temperature.
• 'Frosting' is a particularly difficult fault to avoid with honey that granulates rapidly. It is caused by the air bubbles present in the honey. When placed too soon into a cold store the honey shrinks away from the shoulders and neck of the honey jar forming white irregular patterns. As stated before, honey that granulates slowly or granulates at higher temperatures, greater than 57°F (14° C) seldom exhibits this fault.
• Fine granulation minimises the water content between the crystals and thereby minimises fermentation.
• A vertical streak in granulated honey with the surface texture rough indicates fermentation. At an advanced stage fermenting honey has a characteristically unpleasant smell.

6.10 Describe the spoilage of honey particularly by fermentation (including the effect of water content, storage temperature and the presence of yeast).

6.10.1 Honey can be spoilt by:

• The presence of foreign particles carelessly added during extracting or straining eg. small amounts of oil used to lubricate the rotating parts of the extractor or the draining tap, dust particles etc.

• Unpleasant odours that have been absorbed during storage or extraction. All storage buckets must be absolutely clean and free from any odour. All lids must fit to prevent any air being absorbed ie. air-tight.
• Poor presentation eg. unclean jars, lids and sticky containers.
• Misuse of medicaments by treating bees at the wrong time of the year ie. whilst the supers are in place.
• Feeding of sugar syrup to the bees at inappropriate times so that the sugar syrup is stored with the floral honey.
• Illegally mixing high fructose corn syrup to the honey. HMF is formed during the process of producing invert sugar from sucrose by acid hydrolysis. A high level of HMF can be an indication of adulteration with invert sugar.
• Fermentation of the honey.

6.10.2 Water content of honey.

Before removing any honey from the hive for extraction the beekeeper must first make sure that the honey is 'ripe'. Honey sealed in cells is considered ripe and suitable for extraction but any unsealed cells should first be tested by holding the comb horizontally over the top bars of the open stock and vigorously shaking the comb. Should any nectar fall onto the top bars from the comb then the 'honey' is considered unripe. These combs of unripe 'honey' should remain on the hive for the bees or until the cells have been sealed over by the bees. Once the honey is extracted, the specific gravity of the honey can be taken using a hydrometer or an easier method is to use a refractometer. With this last method honey could be tested before extraction (ie. taking a sample directly from a sealed cell), to test samples of honey at a honey show or prior to storing or bottling. A correction scale is supplied with the instrument to adjust the reading for the ambient temperature. The scale is calibrated to read correctly for 68°F (20° C) eg. for 59°F (15° C) deduct 0.29 from the % of sugar or at 69.8°F (21° C) add 0.06 to the % sugar. This instrument can also be used for testing the sugar content of solid substances such as apples and grapes. It should be noted that the water content of honey affects the density and viscosity of honey.

6.10.3 Fermentation of honey.

All honeys contain yeasts derived from the bees or the flowers. These yeasts are osmophilic yeasts ie. tolerant to high concentrations of sugar. Unripe honey in the hive will ferment and cause dysentery in the bees if not processed and sealed with wax in the cells. Fermentation in extracted honey is associated with the granulation process. Any small particle left in the honey after straining will provide a nucleus on which the sugar crystals form. When honey crystallises, water molecules are released into the ungranulated honey increasing the water level. The chemical changes which take place when fermentation occurs can be expressed as:

$$C_6H_{12}O_6 \ + \ YEAST \ = \ 2CO_2 \ + \ 2C_2H_5OH$$

6.10.4 The risk of fermentation is increased by a high moisture and yeast count:

Moisture level	Yeast count
<17.1 %	No fermentation regardless of yeast count
17.1-18%	Safe if yeast count < 1,000/g
18.1 - 19%	Safe if yeast count <10/g
19.1 - 20%	Safe if yeast count < 1/g
> 20%	Always in danger.

• Numerous osmophilic yeasts have been isolated in honey including *Zygosaccharomyces spp*, *Torula mellis, Schwanniomyces occidentilis, Schizosaccharomyces occidentilis, Saccharomyces spp. etc.*

• To destroy yeasts, honey is heated to 145ºF (63º C) for 30 minutes or raise to 160ºF (71º C) for 1 minute followed by rapid cooling. This will prevent fermentation occurring.

• If fermentation is established then the honey must be heated to a temperature of 200ºF (94ºC) to destroy the yeasts. However, such honey can only be sold as cooking honey after heating to this extent.

• Heating honey destroys the subtle aroma, darkens the honey, increases the HMF and decreases the diastase activity.

• Signs of honey fermenting are:

Sour flavour and unpleasant aroma due to the presence of ethanoic acid (vinegar).
Streaks appearing at the side of the jar. This can be confused with the white mottling at the neck of the jar due to frosting.
A layer of air bubbles on the surface of the honey caused by the presence of CO_2.
A heaving surface.

6.10.5 Storing honey at high temperatures.
Any increase in the storage temperature above 50ºF (10ºC), such as happens in the spring, will help to initiate the fermentation within the storage buckets. Pressure will build up inside the buckets. Fermentation also occurs in bottled honey. Bottled honey prepared as 'set' honey should be sealed with an air tight lid, stored at 57ºF (14ºC) until the crystallisation is complete. Then store below 50ºF (10ºC) to prevent any fermentation taking place.

** ** ** **

STINGS

The Candidate will be able to:

7.1 Describe how to deal with a person who has been stung by a bee but shows no effect other than discomfort and slight local swelling.

7.1.1 General.

It is almost impossible to practise the craft of beekeeping without being stung. Now and again an innocent bystander may receive a sting even when well protected by a veil and bee suit. Prospective beekeepers who have a history of asthma, allergic reactions or other constitutional weaknesses should always take advice from their G.P. before taking up the craft. For some beekeepers anti-inflammatory or antihistamine drugs such as aspirin, Piriton or Triludon may be prescribed. Children are usually very upset when stung and demand lots of attention and reassurance. The moral is to keep docile bees, only allow interested persons near hives of bees when the weather is fine and the bees are busy foraging for nectar. Any attendant in the apiary should be protected by bee veil and suit or similar attire.

7.1.2 The stinging worker bee.

Only the mature worker bee, over 14 days old, is capable of stinging. It is a defensive response stimulated by any alien creature attempting to disturb the hive. Bees are very sensitive to vibration and strange odours. Always avoid any jarring of the hive or frames and never use after-shave lotion, hand cream, hair spray or perfume when attending to bees. After stinging, the bee will die, leaving behind the stinging apparatus including the 7th abdominal ganglion (the last bundle of nerves). Young house bees, up to about 14 days and drones do not sting. Queens only sting other queens.

7.1.3 The effect of stings.

• Most beekeepers at the first sting experience discomfort, pain, reddening of the skin and swelling. The severity seems to vary depending on the site and the number of stings. But over the years a natural immunity will be built up and the discomfort and swelling will be minimal.
• Later extensive swelling or an irritating rash may occur taking 12 hours to reach its maximum and 2 or 3 days to resolve. These symptoms may indicate an increasing sensitivity to bee venom eg. a sting on the hand may result in swelling of the fingers, wrist and forearm.
• A generalised reaction with symptoms of difficulty in breathing, skin rash, palpitations, vomiting and faintness occurring within minutes of a sting indicates a severe reaction (anaphylaxis) requiring emergency medical attention. To be dealt with in section 7.2.

7.1.4 First aid treatment.

• The barb should be removed as soon as possible by a scraping action of hive tool, finger nail, knife or similar flat ended tool. It is thought that pulling the barb out by first finger and thumb

241

method involves squeezing the poison sac and injecting more venom. Quick action in removing the sting will reduce the amount of venom injected and lessen any reaction.

• Move away from the stock of bees as the pheromones given off by the glands of the sting apparatus will attract other bees.

• A first sting is always painful, especially to a child. A reassuring, sympathetic attitude accompanied by the application of ice or calamine lotion will soothe the swollen area. Dry the tears, give a sweet drink and instruct whoever is on standby to watch for any other signs of distress.

• Adults too need reassuring, novice beekeepers can become quite upset if the sting is on the scalp, ear, tip of the finger or anywhere where there is little spare skin to expand. Always enquire after the victim's general health and advise him/her to seek medical advice if any further symptoms appear.

•A sting on the face may cause further swelling during the following hours. These stings always seem more painful, if the site of the sting is on the mouth, neck, near the eye or on the eyeball medical advice should be sought ASAP.

• A seasoned beekeeper will just remove the sting, smoke the affected area in order to disperse the alarm pheromones present and proceed with the task in hand.

7.1.5 Other relevant points.

• Bees kept adjacent to neighbours must be beyond reproach as far as temper is concerned and they must not be of a swarming tendency. Always consider your neighbours; bees must not cause nuisance by stinging, swarming or soiling the washing; all are unacceptable.

• In all the years we have kept bees in our garden, only once has a neighbour been stung. She was about 6 feet from one of our hives weeding her garden the other side of an open mesh fence. She said it didn't matter but we told her it did matter to us and the offending stock was requeened although it was not, in our opinion, a bad stock.

• It is important to manipulate your bees in a garden without gloves if you are overlooked by neighbours, it tells them that the bees are unlikely to be a menace to them. It is even better if the bees can be kept out of sight from neighbours but always ensure that you approach the hives without a veil and similarly when leaving the bees.

• It is good practice never to open up or manipulate a colony in a garden apiary when the next door neighbours are using their garden.

• Good public relations and bees are just plain common sense but do ensure that insurance is kept valid for third party liability. Litigation seems to be the order of the day and attitudes can change dramatically if a serious incident occurs.

• Note the requirements for section 1.2 of the syllabus.

• To summarise, stings can be minimised as follows:

    a) Beekeepers maintaining stocks of docile bees by culling the queens producing nervous bees and those with strong defensive traits eg. followers.

    b) Handling colonies correctly i.e. no jarring of frames, no fast movements, no squashing of bees or banging of hives.

    c) Only open stocks under good weather conditions. Thunderstorms or approaching rain clouds definitely affect the temper of the bee.

    d) Always wear a veil to protect the eyes, the nose, the mouth and ears where there is a proliferation of mast cells in the skin, (mast cells are pharmacological 'time bombs'; when they rupture they release powerful chemicals that can effect various tissues nearby, such as blood vessels or smooth muscle).

e) Use protective clothing of correct material eg. cotton. It is said that woollen garments or garments dyed blue are best avoided. In our opinion the temperament of the bee is the predominant factor and not the clothing.

f) Dispersing the sting pheromone by the application of the smoker to the site of injection does discourage other guard bees being attracted to the site of the first sting. For stings on the hands, place momentarily a hot part of the smoker on the sting area to evaporate as quickly as possible the volatile pheromone and then smoke the area to mask any remaining smell.

g) Refrain from using perfumes, aftershave lotions, hair shampoos, nail varnish, hair sprays and other similar cosmetics prior to working with bees as these have been known to elicit a stinging reaction from bees due to the similarity in the chemical make up to the isopentyl acetate in the alarm pheromone of the sting chamber.

h) Site stocks as far away as possible from the general public.

i) To the non-beekeeper a sympathetic attitude will often solve the immediate problem but should there be any severe reaction, as may be the result of a sting close to or on the eye ball or in or around the mouth or neck, then it is always safest to obtain medical advice.

7.1.6 Other points of interest.

• Most beekeepers expect to receive the odd sting and take no further action beyond removing the barb.

• About 15% of the population have an atopic constitution and roughly 50% of the children of two atopic parents will be similarly afflicted. This group includes individuals with a personal history of hay fever, eczema, asthma, allergic rhinitis and urticaria. They may show progressive worsening in their reactions to stings developing general symptoms such as nausea, skin rashes and respiratory difficulties. Medical aid should be sought immediately should anyone have difficulty in breathing. Taking non-inflammatory drugs eg. aspirin or Piriton under medical advice one or two hours before working in the apiary does reduce the reaction to stings. It should be remembered that antihistamines cause drowsiness so that driving the car is unwise after this kind of medication. Antihistamine creams if used continually can cause dermatitis, if swelling is extensive and itchy a steroid cream may be prescribed by a G.P.

• For the hypersensitive person who wishes to continue beekeeping or for the members of the beekeeper's family who are exposed to the danger of being stung and are hypersensitive, a course of immunotherapy can be arranged through their family doctor.

•Any sign of a generalised reaction as a result of a bee sting should be treated with care and medical advice sought.

7.2 Describe precisely the action to take when a person who has been stung by a bee, exhibits a severe reaction or anaphylactic shock.

7.2.1 Definition of anaphylaxis.

A generalised reaction with signs and symptoms of difficulty in breathing, skin rash, palpitations, confusion, vomiting, faintness caused by a falling blood pressure. All or some of these signs and symptoms occurring within minutes of a sting indicates a severe reaction (anaphylaxis) requiring emergency medical attention. **Do not attempt to take the patient to hospital, you may get caught in a traffic jam. Call an ambulance which is equipped with a siren and flashing lights to speed through the traffic. Time is of the essence. A mobile phone is a great asset in these circumstances.**

7.2.2 First aid treatment while waiting for the arrival of the paramedics.

1. Move the patient well away from the bees.
2. Give an injection of epipen (adrenaline) if this is carried by the patient.
3. Remove any dentures.
4. Keep the airway clear.
5. Release any tight clothing especially around the neck.
6. Lay the patient in the recovery position. **Please see Appendix 9**.
7. Keep the patient warm.
8. Stay with the patient at all times. Death can occur due to circulatory failure or airway obstruction.

** ** ** **

APPENDIX 1 - PROSPECTUS

Aims

To encourage all new and many existing beekeepers to improve their understanding and practice of beekeeping.

The Candidate will be able to demonstrate a broad range of beekeeping skills and adequate understanding. The Candidate will be required to show that their management of colonies is aimed at producing a strong healthy force of foraging bees for the expected honey flows and that the resultant crop is processed for consumption in a hygienic manner with due regard to legal requirements.

Conditions of Entry

The Candidate shall have been awarded the BBKA Basic Certificate or an equivalent certificate acceptable to the Board. The date when this certificate was awarded shall be entered on the Application Form.

The Candidate shall have kept and managed bees for at least three years. A Competent Person who is familiar with the Candidate's beekeeping shall sign a statement to this effect, on the Application Form.

The Secretary to the Board shall have received a completed Application Form and fee by the 28th February in the year the Candidate intends to be assessed.

The Assessment

The Assessment will normally be conducted in May, June or July at the Candidate's apiary and will last about two hours. The Assessor will visit the Candidate's apiary and test his/her practical skills and knowledge of the important aspects of beekeeping and bee products, as defined in the Syllabus. The Assessor will observe the Candidate's practical skills at opening and manipulating colonies and note the correct use of beekeeping equipment and apiary hygiene. Honey preparation facilities will be inspected and the Candidate will describe their procedures for extracting and packaging honey. The Candidate's understanding of beekeeping, as covered in the Syllabus will be assessed through a discussion with the Assessor who may use the Candidate's Record Book as the basis of the assessment. An Assessor appointed by the Board shall conduct the Assessment except for a retake when there will normally be two Assessors. The Board may wish a trainee Assessor or Board member to be present as an observer, but prior written agreement of the Candidate shall be obtained.

Preparation

The Candidate will prepare their apiary to show their approach to general beekeeping, queen rearing and swarm control. The Candidate may decide to attend a series of lectures, join a study group or follow private study using standard texts to prepare for the Assessment. Suggested texts and specialist leaflets are listed after the Syllabus. In addition, the Candidate will be required to have kept a Record Book of activities in the apiary for at least a season. Guidance notes on Maintaining Beekeeping Records are available from the BBKA and are recommended for all beekeepers even if they are not taking this Assessment.

Equipment

The Candidate will ensure the following equipment is ready prior to the Assessment, for inspection by the Assessor, and ready for immediate use:

Three Honey Production Colonies of honeybees.
One Nucleus Colony of honeybees.

Sufficient spare equipment:	for preparing colonies for moving elsewhere;
	for feeding the colonies;
	to produce an artificial swarm;
	for queen introduction;
	for queen marking and clipping;
	for swarm collection.

Honey and wax processing and packing equipment.
Samples of liquid and set honey labelled as for sale (a minimum of 6 jars).
Blocks of wax suitable for sale (minimum weight 25 g).
Personal protective equipment.

Procedure

The Assessor will observe the Candidate's practical skills at opening and manipulating colonies and note the correct use of beekeeping equipment and apiary hygiene. Honey preparation facilities will be inspected and the Candidate will describe their approach to extracting and packaging honey. The Candidate will also be assessed on understanding of beekeeping as covered in the Syllabus. The Assessor may decide to use the Candidate's record book to lead the discussion on beekeeping. Assessors will use standardised assessment sheets supplied by the Examinations Board so that the outcome may be moderated if necessary.

Award of Certificate

There are no grades given. The Candidate either reaches the required standard or not. The Candidate will normally be informed of the outcome within two weeks of the Assessment. The Certificate in Beekeeping Husbandry will be sent to the Local Examination Secretary for presentation to the successful candidates at a later date.

** ** ** **

APPENDIX 2 - RECORD BOOK

The Record Book may be kept in any convenient form by the Beekeeper and will be used to record the activities and conditions found on each and every inspection of a specific hive. It will also provide information on the performance of the apiary including the quantity of honey taken and processed from the hives.

The Record Book will provide a continuous record for at least one season.

There will be an entry in the Record Book each time the beekeeper visits the apiary and manipulates a colony.

The beekeeper will enter the date and time of the visit and for each colony and will use a quantitative method of assessing the following attributes of the colony:

• The existence of a queen from no evidence to laying queen observed or witnessed.
• The existence or otherwise of queen cells/cups.
• The temper of the colony from very docile to unworkable.
• The degree of disease and probable diseases seen.
• The brood size and pattern.
• Quantity of stores available (including pollen).
• Available space for colony expansion.
• Hive hygiene.

The Record Book will be used to record the activities of the beekeeper, such as:

• Feed.
• Frames/supers added or removed.
• Queen rearing activities.
• Swarm control activities.
• Disease control activities including the use of acaricides.
• Cleaning activities.
• Details of swarms collected.

More information may be found in the BBKA leaflet `Maintaining Beekeeping Records'. An overview of the information required is given in the section on Record Keeping which follows in Appendix 3.

** ** ** **

APPENDIX 3 - RECORD KEEPING

The Record Book is used to give an overview of the Candidate's beekeeping activities and to help plan the work in the season. If it is kept as a Filofax it is possible to insert the hive record cards into the book to provide a complete record of the beekeeping season. However, many beekeepers prefer to keep their hive record cards with their hives. If this is done it is important to keep the cards in the dry and away from the bees otherwise they will chew up the card and all the records will be lost. The Record Book comprises three parts:

• The Apiary Layout

This can be pictorial and shows the location of each colony in the apiary and how they are marked. It is always recommended that hives are marked in some way so that other beekeepers and the public can identify the owners. Talk to your local beekeeping Association about the marking system used in your area. Individual record cards may also be kept with this section.

• Plans for work in the Apiary

This section will hold plans for managing the colonies in the apiary. It is particularly useful to record the activities and timings planned for queen rearing and swarm control. It can also be used as a reminder for repairing hives or buying new equipment. There is no special format for this section but most record keepers find it useful to plan activities using a simple diary approach.

If there is any concern over the general vigour or health of a colony it can be marked here as a reminder to replace certain queens or re-site colonies.

Information will also include the dates when inspections and manipulations are needed to raise new queens and other activities requiring forward planning, such as the preparation needed to provide a colony for an observation hive at a show.

• Records of the season

This will give information on the quantity of honey collected during the season and the quality of the queens. Records will also include the state of the hives and the work needed during the winter months to prepare for the next season.

** ** ** **

APPENDIX 4 - APPLICATION TO ENTER

Application to Enter

These should be made through the Local Examination Secretary of the County Beekeeping Association or directly to the BBKA Examinations Board Secretary at the address given below. Applications are required not later than 28th February in the year the Assessment is to be taken.

Application Form

Any application must be accompanied by a completed Application Form together with the Examination Fee. Cheques should be made payable to BBKA. The dates when relevant certificates were obtained must be entered on the Application Form. Certificates should not be sent.
Ensure that the Certificate of Qualification, on the Application Form, is completed by a Competent Person.

Assessment Fee

The current fee for the Assessment may be obtained from the Local Examination Secretary or the Board Secretary.

<u>AUTHORITY</u>

The above is issued by the BBKA Examinations Board and all communication in respect of this Assessment should be addressed to:

> The Secretary,
> BBKA Examinations Board,
> The British Beekeepers' Association,
> National Agricultural Centre,
> Stoneleigh,
> Kenilworth,
> Warwickshire CV8 2LZ

Published April 2000

** ** ** **

APPENDIX 5 - SUGGESTED READING TEXTS

Practical Beekeeping   Clive de Bruyn
A Guide to Bees and Honey   Ted Hooper
Foul Brood diseases of honey bees   MAFF
Varroa jacobsoni monitoring
and forecasting mite populations   MAFF
So you wish to sell Honey   BBKA leaflet
Bees and Neighbours   BBKA leaflet

Authors' note:-

Leaflets currently available from ADAS are understood to be as follows (received July 2001 from our RBI):

• PB3053 - Foul brood diseases of honey bees: recognition and control.
• PB3054 - Statutory procedures for controlling foul brood.
• PB3611 - Varroa jacobsoni: monitoring and forecasting mite populations within honey bee colonies in Britain. Plus the calculator.
• PB2581 - Managing Varroa.

** ** ** **

APPENDIX 6 - COLONY DEVELOPMENT DURING THE ACTIVE SEASON.

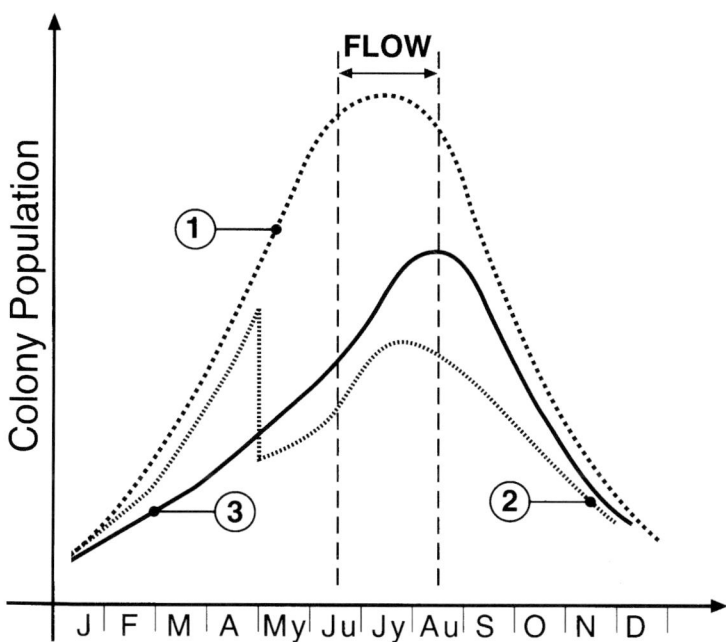

Graph 1. This illustrates a normal healthy colony building up on the spring flow to reach a maximum when the main flow occurs in June/July. The flow stops and the colony population then starts to decrease rapidly due to the queen's reduction in egg laying and the demise of those workers after their three week stint as foragers. The maximum population varies with the efficacy of the queen and the type of bee but in general may be assumed to be about 40,000 bees.

Graph 2. This illustrates a normal healthy colony which swarms in about April/May and then proceeds to requeen itself. It will be clear that it has a very much reduced colony population, and hence foraging force, when the main flow starts and therefore resulting in a very reduced honey crop. The amount of honey collected is directly proportional to the colony population.

Graph 3. This illustrates the build up of a diseased colony, say one with Nosema. It has a very reduced population and reaches its peak as the main flow is ending. Again this has a very serious effect on the amount of honey collected by the colony.

Considering the three graphs shown above it will be clear that in order to ensure the maximum honey crop it is necessary to have a disease-free colony which builds up normally on the spring flow and does not swarm. It also illustrates the necessity of checking the colony for disease in the spring. If the colony is declared disease-free and its performance is akin to Graph 3, then it is more than likely that the queen is failing and needs replacement.

** ** ** **

251

APPENDIX 7 - MIGRATION AND EVOLUTION OF THE HONEYBEE.

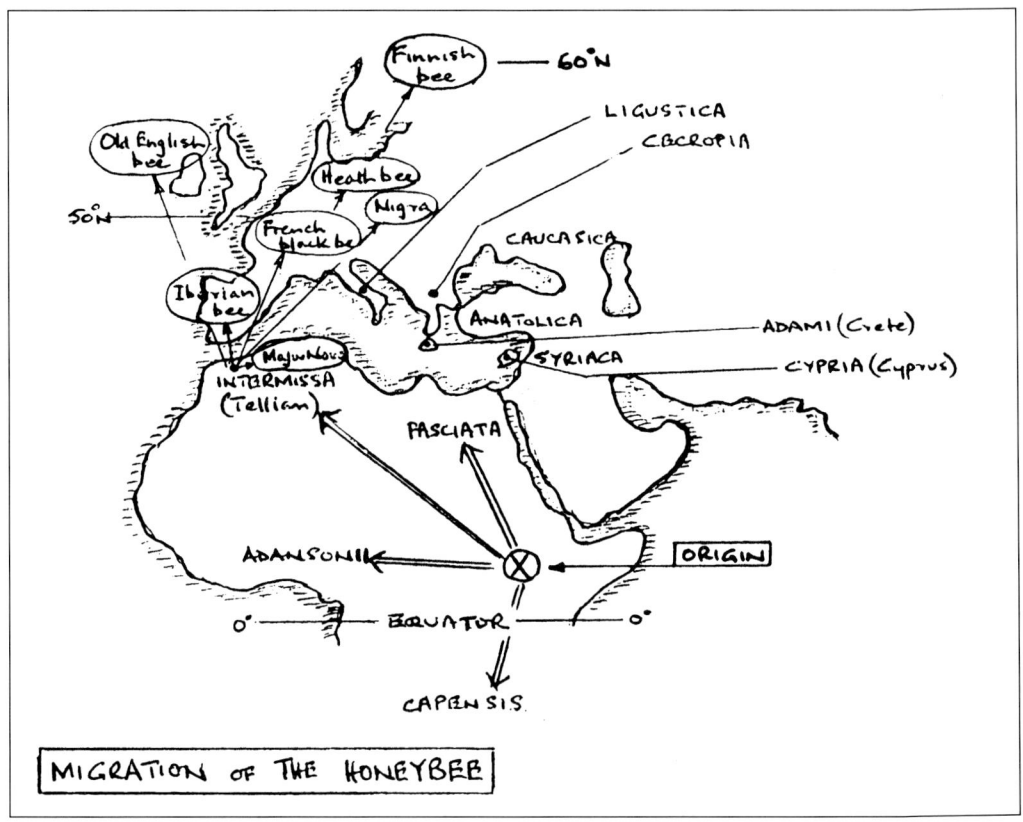

MIGRATION OF THE HONEYBEE

The above diagram shows the migration of the honeybee, from its origin in what is now Kenya, during the last 30 million years. All the major races shown by double arrows may be considered primary races and the one of importance to the UK situation is the *Apis mellifera intermissa*. It is from this bee that migrated northward into north-west Europe that all the other sub-species have evolved, such as the French Black Bee and the Old English Bee shown with single arrows emanating from north west Africa on the diagram, ie. from the area of Morocco, Tunisia and Algeria. It migrated as far north as latitude 60° close to the Arctic circle and became the Finnish Bee, a bee that is characterised by its ability to tolerate long periods of confinement in very cold conditions without cleansing flights. This is a truly remarkable evolution from an ancestor derived from the tropics.

The western honeybee, *Apis mellifera,* was transported to America initially by the pilgrims and to Australia and New Zealand by the settlers to these Antipodean countries. Until man took a hand in these migrations there were no honeybees in either of the two regions. The original bees were taken in skeps on the old sailing ships with a voyage time of about 6 months to the Antipodes; quite remarkable and a topic that would be an interesting line of research into the whys and wherefores of these exploits.

** ** ** **

APPENDIX 8 - AVERAGE COLONY POPULATION CYCLE

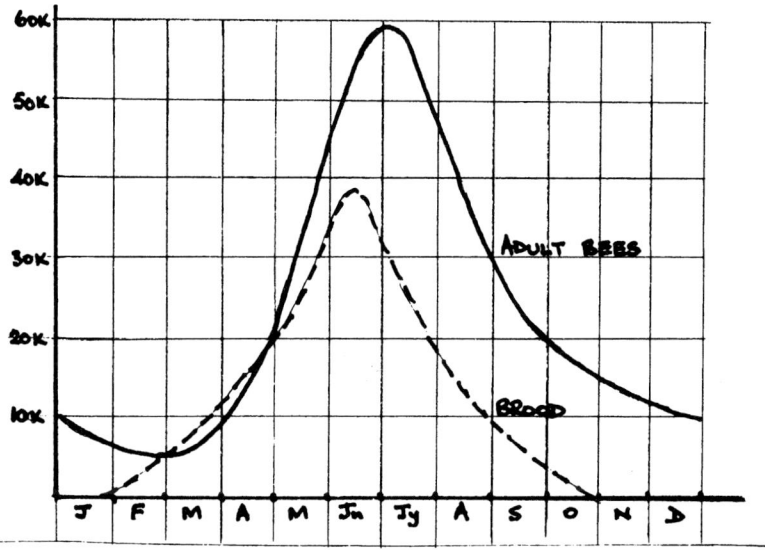

• Brood = adult bees twice per year.
• Brood > adult bees from February to April. This is a very critical time in the annual colony cycle (nb. the danger of chilling brood and not having enough adult bees to incubate this brood).
• Brood peaks in early/mid June.
• Adult bees peak end June/start July (3 weeks after brood peak); this is the time the main flow usually starts when the maximum foraging force is required.
• After the main flow the population starts to decrease, rapidly at first (old foragers dying off) then more slowly as the winter bees (6 month life) start to appear in the colony.
• The minimum adult population occurs c. end February (c. 5000).
• The maximum population will vary from 40,000 to 60,000 depending on the fecundity and strain of the queen.
• The population builds up on the 'spring flow' often using all the income and storing very little.
• The maximum adult population stores very large amounts in a short time for winter (much less brood to care for).
• The reduced population allows adequate reserves for winter.
• Brood rearing ceases in late autumn and starts again after the winter solstice when the days start to lengthen.
• There is a continual decrease in population throughout the winter so dying bees are not abnormal at this time. The healthy colony removes any that die in the hive.

It should be noted that the graph is a representation of average conditions and the local flora and climatic conditions will modify it accordingly. Similarly, these local variations mean a peaky graph and not a smooth curve as shown.

** ** ** ** **

APPENDIX 9 - THE RECOVERY POSITION

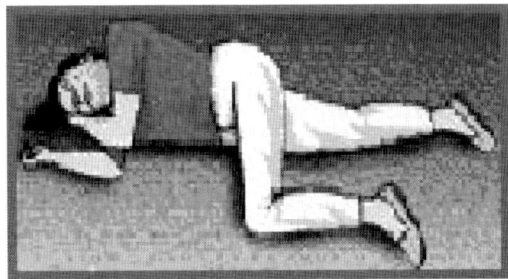

The recovery position is the correct position in which to place a victim **who is breathing**, while waiting for help to arrive. Do not put a person in the recovery position if you suspect that he has a spinal injury or severe fractures.

Lie the victim on his back and kneel beside him. Tilt the head back and lift the chin to open the airway.

If he is wearing glasses, remove them; loosen his shirt, belt and pants.

Straighten both legs and place the arm nearest to you at right angles to the victim's body, elbow bent, with the palm of the hand uppermost.

Bring the far arm across his chest and hold the hand against the victim's cheek, palm outwards. With the other hand, grasp the furthermost thigh and pull the knee up, keeping the foot on the ground.

Support the victim's head by keeping the hand pressed against his cheek with one hand. Roll him towards you with the other hand holding the bent knee of the farthermost leg.

Tilt the head back to open the airway, adjust the hand to support the head.

Adjust the uppermost leg so that the hip and knee are at right angles.

Check that the victim's breathing and pulse are regular.

** ** ** **

INDEX

259

** ** ** **